SpringerBriefs in Computer Science

Series Editors

Stan Zdonik
Peng Ning
Shashi Shekhar
Jonathan Katz
Xindong Wu
Lakhmi C. Jain
David Padua
Xuemin Shen
Borko Furht
V. S. Subrahmanian
Martial Hebert
Katsushi Ikeuchi
Bruno Siciliano

For further volumes:
http://www.springer.com/series/10028

Xiaofeng Tao · Qimei Cui
Xiaodong Xu · Ping Zhang

Group Cell Architecture for Cooperative Communications

 Springer

Xiaofeng Tao
Key Laboratory of Universal Wireless
 Communications
Beijing University of Posts and
 Telecommunications
No. 10 Xitucheng Road
Haidian District
100876 Beijing
People's Republic of China

Xiaodong Xu
Key Laboratory of Universal Wireless
 Communications
Beijing University of Posts and
 Telecommunications
No. 10 Xitucheng Road
Haidian District
100876 Beijing
People's Republic of China

Qimei Cui
Key Laboratory of Universal Wireless
 Communications
Beijing University of Posts and
 Telecommunications
No. 10 Xitucheng Road
Haidian District
100876 Beijing
People's Republic of China

Ping Zhang
Key Laboratory of Universal Wireless
 Communications
Beijing University of Posts and
 Telecommunications
No. 10 Xitucheng Road
Haidian District
100876 Beijing
People's Republic of China

ISSN 2191-5768   ISSN 2191-5776 (electronic)
ISBN 978-1-4614-4318-6   ISBN 978-1-4614-4319-3 (eBook)
DOI 10.1007/978-1-4614-4319-3
Springer New York Heidelberg Dordrecht London

Library of Congress Control Number: 2012935252

© The Author(s) 2012
This work is subject to copyright. All rights are reserved by the Publisher, whether the whole or part of the material is concerned, specifically the rights of translation, reprinting, reuse of illustrations, recitation, broadcasting, reproduction on microfilms or in any other physical way, and transmission or information storage and retrieval, electronic adaptation, computer software, or by similar or dissimilar methodology now known or hereafter developed. Exempted from this legal reservation are brief excerpts in connection with reviews or scholarly analysis or material supplied specifically for the purpose of being entered and executed on a computer system, for exclusive use by the purchaser of the work. Duplication of this publication or parts thereof is permitted only under the provisions of the Copyright Law of the Publisher's location, in its current version, and permission for use must always be obtained from Springer. Permissions for use may be obtained through RightsLink at the Copyright Clearance Center. Violations are liable to prosecution under the respective Copyright Law.
The use of general descriptive names, registered names, trademarks, service marks, etc. in this publication does not imply, even in the absence of a specific statement, that such names are exempt from the relevant protective laws and regulations and therefore free for general use.
While the advice and information in this book are believed to be true and accurate at the date of publication, neither the authors nor the editors nor the publisher can accept any legal responsibility for any errors or omissions that may be made. The publisher makes no warranty, express or implied, with respect to the material contained herein.

Printed on acid-free paper

Springer is part of Springer Science+Business Media (www.springer.com)

Foreword

For the past decades, cellular network architectures have achieved great success during the development of mobile communications. However, today cellular network architectures encounter some unprecedented challenges, which include higher data rate, higher carrier frequencies, smaller coverage area, and stronger BS processing abilities. What changes will occur in future mobile network architecture? How to improve the network architecture?

In 2001, Prof. Xiaofeng Tao et al. proposed the Group Cell architecture for the first time in order to solve the mentioned challenges. The Group Cell architecture is based on distributed cellular architecture with multi-antenna techniques for cooperative communications, which can increase the coverage area and employ the advantages of multi-antennas techniques. Group Cell architecture also has an enhanced flat (not traditional hierarchy) network architecture for IP services. Moreover, Group Cell has covered the scenarios of Coordinated Multiple-Point Transmission and Reception (CoMP), which was proposed in April, 2008 for 3GPP LTE-Advanced as a tool to improve the cell-edge throughput and increase system capacity.

Since 2001, the authors had performed plenty of research about Group Cell for cooperative communication. Based on the Group Cell architecture, the authors studied capacity analysis, slide handover and power allocation in Group Cell, etc. Furthermore, the authors developed the first 4G TDD Trial system for Group Cell and completed the validation of the above key technologies in 2006. Up-to-date, based on the investigations of Group Cell, they have made some good achievements including publications and patents. In this book, the authors share their main findings.

With a balanced blend of theoretical analysis and system verification, this book provides researchers and practitioners with basic properties of Group Cell architecture for cooperative communication and system development. All the techniques introduced in this book are quite new. In addition, this book makes an easy-to-follow presentation and is ideal for graduate students.

Hequan Wu

Hequan Wu
Academician, Chinese Academy of Engineering

Preface

The reason for writing this book started 11 years ago, as a result of an innovative idea on the cell architecture in mobile communication network. At that time, there was a growing tendency for high speed wired and wireless data rates. The data transfer bottleneck occured at the interface between Radio Access Network (RAN) and the Core Network (CN) as they converged at the Base Station Controller (BSC) via different Base Stations (BSs). An ideal solution is to bypass the BSC and connect the BSs directly with the CN avoiding the possible bottleneck. Control entities resided in traditional BSCs are directly incorporated to the CN, which resulted in a prototype of LTE network structure of nowadays.

Meanwhile, future mobile communication systems will continue to have higher carrier frequencies, smaller coverage area, and stronger BS processing abilities. Thus, the future scenario that multiple distributed AUs (antenna units, antenna elements / antenna arrays) are connected to the same BS would likely be realized soon. We first proposed the Cooperative Communication Oriented Group Cell Architecture in 2001, by combining the technologies (available at that time) such as transmit diversity, micro cells, joint transmission, and MIMO.

The fundamental idea of Group Cell is to convert the traditional hierarchy RAN architecture into an enhanced flat one in which BS will access the CN directly, without the support of BSC. Group Cell is characterized by several adjacent cells in order to serve a certain MT (mobile terminal) in an efficient way. In Group Cell, the physical layer technology greatly affects the way signals are transmitted. For instance, using distributed antenna technique, the same signals can be transmitted by different transmitting AUs, whereas in case of joint transmission, each transmitting AU may send different signals. Group Cell can be formed by adaptive AU selection at physical layer, in which AUs can be selected from both intra BSs and inter BSs, or even from heterogeneous BSs. Moreover, contrary to the traditional static cells, Group Cells are not stationary anymore. They can move together with the moving MT, thus MT will always be located in the center of a Group Cell. By employing this, cell edge problem can be avoided and the handover probability can also be reduced significantly. This also constitutes a prototype of CoMP in 4G networks.

The trial systems with Group Cell Architecture for Cooperative Communications were implemented under the support of the Chinese NSFC and MOST 863 high-tech program in 2006. To the best of the authors' knowledge, these were the first trial systems ever attempted that included features such as 4G TDD, CoMP, mobile Internet with IPV4/IPv6. The 4G TDD and CoMP features were never implemented before. The integration of 4G TDD and IPv6 was also a first attempt of its kind. The 4G TDD feature also supported multi-BS multi-user handover in multi-cell, single-frequency networking,high vehicular speed, and low BER (less than 10^{-8}). The CoMP feature supported multi AUs cooperation in transmit/ receive signals. The standardization of technologies mentioned above is in progress, some of which has already been standardized.

In 2009, with lots of preliminary works on cooperative communication in Group Cell, the author carried out further research in Stanford University. After analyzing the capacity of single user, multi-user diversity and ergodic capacity in Group Cell, the author proposed the Slide Handover and fast group cell selection scheme, and evaluated their performances. The authors also studied the energy conservation problem in Group Cell, including ICIC-based downlink resource allocation and downlink power allocation for capacity maximization.

In 2011, under the support of *Prof. Sherman Shen* and *Springer Press*, we decided to publish our research work carried out during the past 10 years. All comments and suggestions for improvements of this book are welcome.

Acknowledgments

The authors would like to acknowledge the support of the National Natural Science Foundation of China (NSFC), the 863 program, and the International Science & Technology Collaboration program from Ministry of Science and Technology (MOST), Ministry of Industry and Information Technology (MIIT), and Science & Technology Committee of Beijing. Most of the work in this book is sponsored by the NSFC projects (60496312, 61027003, 61001116, 61001119), 863 projects (2003AA12331004, 2006AA01Z260, 2006AA01Z283), National Science and Technology Major Project 2009ZX03003-011-02, International Science & Technology Collaboration projects 2008DFA12110, International Scientific and Technological Cooperation Program 2010DFA11060, China-EU International Scientific and Technological Cooperation Program 0902 and project D08080100620802 from Science & Technology Committee of Beijing.

The authors would also like to thank the following for their significant contributions in the ideas and implementation of Chinese 4G TDD Group Cell Trial network: Prof. Xiaohu You, Prof. Guo Wei, Prof. Shaoqian Li, Prof. Jing Wang, Prof. Guangxi Zhu, Prof. Youyun Xu, Prof. Youxi Tang, Prof. Xiqi Gao, Dr. Yong Wang, Dr. Qiang Wang, Dr. Liang Qian, Dr. Hao Liu, Dr. Jian Liu, Dr. Gang Su, Dr. Jin Xu, Dr. Mingyu Zhou, Dr. Wenjun Wang, Ms. Jing Shu, Ms. Xiaoting Mu, et al.

The authors are thankful to Ph.D candidates: Mr. Waheed ur Rehman, Ms. Xiangling Li, and Master candidates: Ms. Yingyue Xu, Mr. Chengjin Luo, Ms. Ting Fu for their contribution of editing work and figure reproduction.

Special thanks to Prof. Sherman Shen, he made this book possible.

Contents

1 Introduction to Group Cell Architecture 1
  1.1 Introduction 1
    1.1.1 Relay 2
    1.1.2 DAS 2
    1.1.3 Multicell Coordination 3
  1.2 Group Cell Architecture 4
  1.3 Typical Scenarios of Group Cell Architecture 7
  1.4 Coordinated Multiple-Point Transmission and Reception 8
  1.5 Trial Systems for Cooperative Communications 9
  1.6 Outline 11

2 Capacity Analysis 15
  2.1 Capacity Analysis of Single-User in Group Cell 15
    2.1.1 Shannon Capacity Analyses of Group Cell 15
    2.1.2 Outage Capacity Analysis of Group Cell 17
  2.2 Capacity Analysis with Multi-User Diversity in Group Cell 19
    2.2.1 Calculation of Capacity Gain with MUD 21
    2.2.2 Performance Analysis and Evaluation 24
  2.3 Ergodic Capacity of Group Cell Systems
    with Power Constraints 24
    2.3.1 System Model and Problem Formulation 26
    2.3.2 Capacity Analysis for Group Cell
    with Power Constraints 27
    2.3.3 Performance Analysis and Evaluation 31
  2.4 Summary 33

3 Slide Handover 35
  3.1 Slide Handover Mode 35
    3.1.1 Group Cell and Slide Handover 35
    3.1.2 Performance Analysis and Evaluation 37

  3.2   Fast Cell Group Selection Scheme Mode   39
         3.2.1   System Scheme.   39
         3.2.2   Performance Analysis and Evaluation   43
  3.3   Summary   46

4   Power Allocation of Group Cell System   47
  4.1   Downlink Resource Scheduling for ICIC   47
         4.1.1   GIR-Based Subcarrier Resource Allocation   48
         4.1.2   Balanced SIR-Based Power Allocation   49
         4.1.3   Subcarrier Optimization Algorithm   51
         4.1.4   Performance Analysis and Evaluation   51
  4.2   Downlink Power Allocation for Maximizing
         System Capacity   55
         4.2.1   Optimal Transmit Power Allocation   56
         4.2.2   Sub-Optimal Power Allocation Scheme   61
         4.2.3   Performance Analysis and Evaluation   65
  4.3   Summary   66

5   Group Cell Trial Systems.   69
  5.1   Introduction to FuTURE 4G TDD Trial System   69
         5.1.1   Technical Targets   69
         5.1.2   Key Technologies and PHY Link Design.   69
         5.1.3   TDD Frame Structure   70
  5.2   Performance Analysis and Evaluation   71
         5.2.1   Simulation Scenarios.   71
         5.2.2   Performance Analysis and Evaluation   72
  5.3   Trial Equipments and Trial Scenarios   73
         5.3.1   Trial Equipments   74
         5.3.2   Trial Scenario: Campus   74
         5.3.3   Trial Scenario: Highway   77
  5.4   Trial Results of Group Cell Trial System   78
         5.4.1   Point-to-Point Link Performance Trial.   78
         5.4.2   Trial Results for Campus Scenario   80
         5.4.3   Trial Results for Highway Scenario.   83
  5.5   Trial Plan in the Next Phase.   83
  5.6   Summary   87

References   89

Chapter 1
Introduction to Group Cell Architecture

The wireless transmission and networking technologies are the essential components of the mobile communication systems. Due to the recent breakthrough in transmission technologies, cellular communications have entered the era of cooperative communications. The concept of Group Cell architecture was proposed correspondingly by the authors [1, 2].

The Group Cell architecture provides the cooperative communication schemes, some of which were included in coordinated multi-point (CoMP) of 3GPP long term evolution advanced (LTE-A). Moreover, control entities resided in traditional base station controllers (BSCs) are directly incorporated in core network (CN), which resulted in a prototype of LTE network structure of nowadays.

1.1 Introduction

The cellular concept was a major breakthrough in solving the problem of spectral congestion and user capacity. It offered very high capacity in a limited spectrum allocation without any major technological changes. The use of radio channels on the same carrier frequency to cover different areas, which are separated from each other by sufficient distance so as to avoid co-channel interference [3]. Since then we have entered into the era of cellular communication, and the construction of the cellular network has not changed much in the past 30 years.

As the physical layer techniques develop, the data rates of mobile communication services have increased by about 100 times every 6–7 years, and it is predicted in [4] that in 2020, the required data rate will be as large as 100–1000 times the currently served data rate. Next-generation mobile communication systems promise to provide very high data rates and mass wireless access services for broad area coverage. The traditional cellular systems have the following limitations:

X. Tao et al., *Group Cell Architecture for Cooperative Communications*,
SpringerBriefs in Computer Science, DOI: 10.1007/978-1-4614-4319-3_1,
© The Author(s) 2012

Fig. 1.1 Relay system

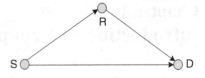

1. Due to the limited transmit power, the transmission distance in traditional cellular network cannot meet the desired high date rate.
2. The frequency of handover in high speed mobile environment (350 km/h) even makes the traditional network harder and much more expensive.
3. What is more, for future mobile communication systems working at frequencies higher than 2 GHz, the cell edge effect becomes more serious due to the larger attenuation of the radio signal.

In order to overcome these problems, some cooperative techniques have emerged, such as relay, distributed antennas systems (DAS), and multicell coordination. All these techniques have turned the traditional cellular system into a cooperative system. Cooperative communications dramatically change the abstraction of a wireless link and offer significant potential advantages for wireless communication networks. In contrast to single-node transmissions, cooperative communications may involve multiple nodes transmitting simultaneously to a receiver. The cooperative communications have the following significant components:

1.1.1 Relay

The basic idea behind cooperative communications can be traced back to the groundbreaking work of Cover and El Gamal on the information theoretic properties of the relay channel [5]. That work analyzed the capacity of the three-node network consisting of a source, destination, and a relay, as depicted in Fig. 1.1. It was assumed that all nodes can operate in the same band, so the system can be decomposed into a broadcast channel (BC) and a multiple access channel (MAC) from the viewpoint of the source and the destination, respectively.

In the year 1998, Andrew Sendonaris, Elza Erkip and Behnaam Aazhang proposed a new form of spatial diversity, in which diversity gain was achieved via the cooperation of mobile users. That is, in each cell, each user had a "partner". Each of the two partners was responsible for transmitting his information as well as the information of his partner, which they received and detected [6]. This idea can be characterized as the specific application of traditional relay model.

1.1.2 DAS

DAS was first introduced to improve the indoor coverage performance of wireless communication systems in 1987 [7]. DAS, in some sense, can also be treated as a

Fig. 1.2 DAS system

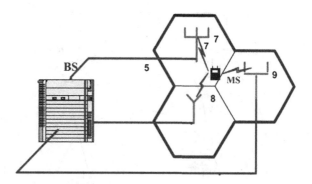

form of cooperative communication systems, in which two or more information sources simultaneously transmit a common message.

Nowadays, for signal quality enhancement and capacity improvement, DAS with multiple remote antenna units (AUs, AEs/antenna arrays) connected to same base station (BS) is gaining more attention as an effective means for providing spatial diversity, as depicted in Fig. 1.2. Unlike typical cooperative relay systems in which the available system bandwidth is shared for transmission to mobile terminal (MT) and for relaying, DAS does not require additional radio resources as remote AUs are typically connected to the BS via dedicated wires such as optical fibers.

1.1.3 Multicell Coordination

In 1991, the "Microcell system for cellular telephone system" (United States Patent 5067147) was proposed to increase the capacity of cellular mobile communication, as depicted in Fig. 1.3. Microcell uses power control to limit the radius of its coverage area, such as a mall, a hotel, or a transportation hub. A cellular telephone system is described as a type wherein a plurality of contiguous cells, each having an assigned set of identification codes, are arranged in such a way to maintain continuous communication with users moving from cell to cell.

In the year 2000, site selection diversity transmission (SSDT) [8] power control, an advanced form of transmission power control was proposed for code division multiple access (CDMA) forward link. SSDT realizes site selection transmission diversity instead of full site transmission diversity used in conventional transmission power control during soft handover (SHO) mode. The major intention of the site selection is to mitigate interference caused by multiple site transmission with conventional transmission power control. Meanwhile the SSDT saves the transmitted power, and also overcomes the power limitation problem in downlink CDMA system.

In contrast to the more conventional forms of single-user space diversity, in 2001, D.TSE [9] built upon a classical relay channel model and examined the

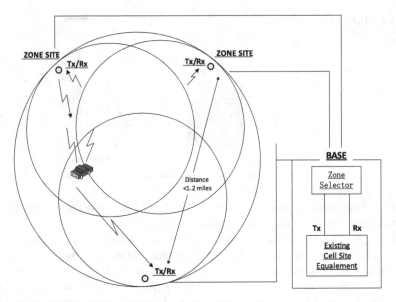

Fig. 1.3 Microcell system for cellular telephone system

problem of creating and exploiting space diversity using a collection of distributed antennas belonging to multiple users, each with their own information to transmit. This form of space diversity was referred to as cooperative diversity because the users share their antennas and other resources to create a "virtual array" through distributed transmission and signal processing.

Apart from the cellular scenario, user cooperation diversity has the potential to be successfully used in wireless ad hoc networks too [10]. The wireless ad hoc network does not contain a fixed infrastructure and a central control unit such as BS.

1.2 Group Cell Architecture

The capacity of the wireless network increases rapidly in recent years. "The wireless capacity has doubled every 30 months over the last 104 years [11]". This translates into an approximately million-fold capacity increase since 1957. Thus, data transfer bottleneck appeared at the interface between the radio access network (RAN) and the core network (CN) as they converged at the BSC via different BSs. An ideal solution is to bypass BSC and connect the BSs directly with CN avoiding the possible bottleneck. Control entities resided in traditional base station controllers (BSCs) are directly incorporated in CN, which resulted in a prototype of LTE network structure of nowadays.

Meanwhile, future mobile communication systems will continue to have higher carrier frequencies, smaller coverage area, and stronger BS processing abilities.

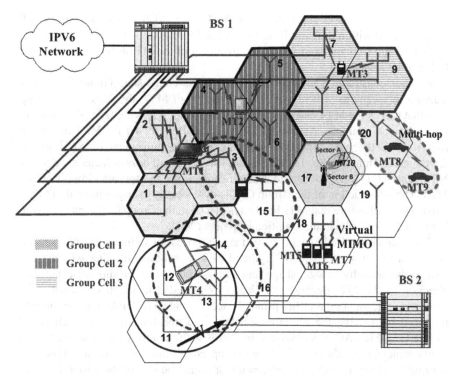

Fig. 1.4 Group cell architecture for cooperative communications

Thus, the future scenario that multiple distributed AUs are connected to the same BS would likely be realized soon. We first proposed the Cooperative Communication Oriented Group Cell Architecture for China 4G TDD mobile communication system [1, 2, 12] in 2001, by combining the technologies (available at that time) such as transmit diversity, micro cells, joint transmission (JT), and multi-input multi-output (MIMO).

A Group Cell is characterized by several adjacent cells that use the same resources to communicate with a specific MT and different resources to communicate with different MTs. The Group Cell Architecture for Cooperative Communications is plotted in Fig. 1.4, which indicates the typical Group Cell configuration. The cells connected to a BS can form one or several Group Cells. The structure, size and topology of the Group Cell are flexible to accommodate different application environments.

In a typical Group Cell based wireless communication system, each BS has several separated AUs. The AU can be single antenna or several antenna elements (AEs). AUs are typically connected to the BS via dedicated wires such as optical fibers as depicted in Fig. 1.4. Actually, the depicted Fig. 1.4 is a hybrid AU with single antenna and several AEs. The function of signal processing is accomplished at the BS. In Fig. 1.4, BS1 has 9 AUs, indexed by $1 \sim 9$. If the size of the Group Cell is 3, we can find 3 Group Cells connected with BS1, e.g. Group Cell 1 with

AU1, AU2, AU3 and Group Cell 2 with AU4, AU5, AU6, and so on. Group Cell can be formed within different BSs. This is fixed Group Cell structure and the AUs of each Group Cell are fixed. With the movement of the MT, different fixed Group Cell can be selected.

Another construction method of Group Cell is called slide Group Cell. Considering the scenario in BS2, with the movement of MT4, the AUs of the Group Cell that serves the MT4 can also update correspondingly with MT4. As shown in Fig. 1.4 as an example, AU11, AU12, AU13 of BS2 served for MT4 in current time-slot. In the next time-slot, with the possible movement of MT4 in the direction shown in the Fig. 1.4, AU12, AU13, AU14 will be re-selected by BS2 for MT4. The AU11 is removed and AU14 is added into the serving Group Cell for MT4. The construction of the Group Cell is dynamically changed instead of being fixed. This will ensure that MTs be always placed in the Group Cell center to avoid both "cell edge" effect and traditional frequent handover.

The slide Group Cell may be viewed as the process of sliding windows. Several cells in a Group Cell could be regarded as in a window and that window can change dynamically in size, shape and slide speed due to the moving speed and direction of the MT. Because the Group Cell needed multicell coordination to serve the user, maximum utility principle access control (MUPAC) algorithm was proposed for solving the group establishing and updating in Group Cell architecture [13]. MUPAC was described in details with the utility function, maximum utility principle and the process of cell group establishing and updating. Based on the quality of service (QoS) requirements of users, the size of the Group Cell can be adaptively adjustable to meet the QoS requirements with the maximum efficiency of antenna and resource.

At physical layer, the signal can be transmitted and received by all the antennas of a group via techniques such as DAS, JT, Space–Time Coding (STC), MIMO, orthogonal frequency division multiplexing (OFDM). It is mentioned that these techniques in Group Cell may be different from traditional DAS, since the transmitted signal from different AU may be different. Therefore, the system's ability to resist interference and system capacity could be improved [14]. In the coverage area of AU18 of BS2, the multi-antenna AU18 can also communicate with single-antenna MT 5, 6, 7 to form virtual MIMO scenario by diversity mode or multiplex mode of MIMO transmission scheme. Multi-hop scenario or relay scenario can also be included in the AU20 of BS2. In the coverage area of AU17 of BS2, two sectors belong to AU17 can also perform the coordinated communication to serve MT10.

As a summary, Group Cell architecture has the notable features as followings:

1. Flat Radio Access Network

   - Without traditional BSC
   - Enhanced flat RAN architecture but traditional hierarchy one
   - All IP-based architecture
   - Short latency

Fig. 1.5 Coordinated communication inside group cell

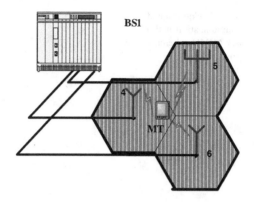

2. Novel Cellular Architecture

   - Slide Handover
   - User always at cell center
   - Solve "smaller cell" problem
   - Avoid "cell edge" effect
   - Avoid frequent handover
   - Enlarge coverage area

3. Fully Explore Space Diversity

   - Distributed Antenna Unit
   - Virtual MIMO
   - Multi-hop, Relay

4. Homogeneous/Heterogeneous network

1.3 Typical Scenarios of Group Cell Architecture

From Fig. 1.4, Group Cell architecture can be seen to have lots of application scenarios. The typical scenarios include coordinated communication inside Group Cell (Fig. 1.5), coordinated communication with multi-user MIMO scheme inside Group Cell (Fig. 1.6), coordinated communication between two BSs (Fig. 1.7), Slide Handover (Fig. 1.8), Fast Cell Group Selection within Group Cell (Fig. 1.9) and so on.

Scenarios in Figs. 1.5, 1.6 and 1.7 will be illustrated with the capacity analyses and power allocation in Chaps. 2 and 4. Scenarios in Figs. 1.8 and 1.9 will be used in Chap. 3 with the Group Cell handover mode research.

Fig. 1.6 Coordinated
communication with multi-
user MIMO scheme inside
group cells

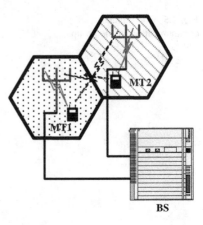

Fig. 1.7 Coordinated
communication between two
BSs

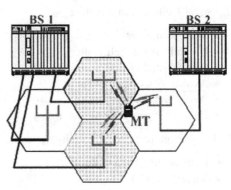

Fig. 1.8 Slide handover
scenario

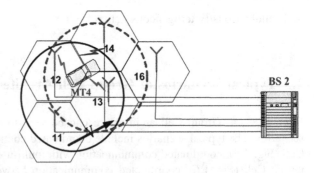

1.4 Coordinated Multiple-Point Transmission and Reception

Based on the concept and architecture of Group Cell, Coordinated Multi-Point
(CoMP) for long term evolution advanced (LTE-A) was proposed in April, 2008
[15], as a tool to improve the coverage of high data rates, the cell-edge throughput
and/or to increase system throughput. The cell concept defined in LTE can also
support the coordinated scenario [16]. CoMP implies dynamic coordination among

Fig. 1.9 Fast cell group selection within group cell

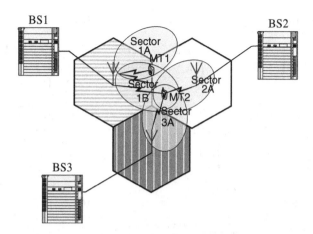

multiple geographically separated transmission points. Examples of coordinated transmission schemes include joint processing (JP) and coordinated scheduling/ beamforming (CS/CB). In JP, data to a single MT is simultaneously transmitted from multiple transmission cells, which can improve the received signal quality and/or actively cancel interference for other MTs. In CS/CB, data is only available at serving cell, but user scheduling/beamforming decisions are made with coordination among cells. In this way, inter-cell interference can be well controlled. Group Cell patented in 2001 has covered all the scenarios of CoMP, which is illustrated in China Patent (ZL 01137188.9, 8/11/2001).

Now, CoMP concept is accepted by 3GPP LTE-Advanced work group, and formally written in 3GPP TR 36.814, which is the technique specification on the further advancements for E-UTRA physical layer aspects. CoMP Release 10 study item (SI) is successfully started in 3GPP RAN 47 meeting on March, 2010. This SI title is the Study on Coordinated multiple point operation for LTE. In Dec. 2010, CoMP Release 11 SI was started again in 3GPP RAN 50th meeting. In future, CoMP will produce larger impact on 4G mobile communication network.

Figure 1.10 shows some scenarios in CoMP, including intra-eNB coordination with SU MIMO, intra-eNB coordination with MU MIMO, inter-eNB coordination with SU MIMO and super cell coordination with MU MIMO. Figure 1.10a, b, and c correspond to the Group Cell scenarios shown in Figs. 1.5, 1.6 and 1.7 respectively. It is clear that the CoMP scenarios are contained in Group Cell.

1.5 Trial Systems for Cooperative Communications

After the Group Cell concept was proposed in the year 2001, the trial systems for cooperative communications based on Group Cell architecture were designed and implemented in 2006 (Fig. 1.11).

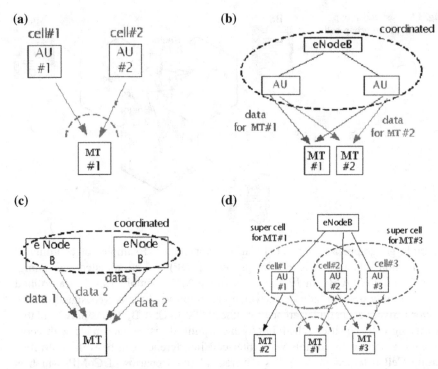

Fig. 1.10 CoMP typical scenarios **a** Intra-eNB coordination with SU MIMO **b** Intra-eNB coordination with MU MIMO.**c** Inter-eNB coordination with SU MIMO **d** Super cell coordination with MU MIMO

The frame design, physical technology and physical-link design, the performance simulation and evaluation, system parameter design and scenario settings were accomplished in 2005. The typical networking environments of the Group Cell system were implemented in Beijing and Shanghai trial fields separately with all the main features of Group Cell architecture, such as Cooperative Communication, adaptive MIMO and soft fractional frequency reuse. Among which, the urban (campus) trial scenarios and the highway trial scenarios were developed and the field tests were on the way.

Then, on Oct. 31st, 2006, Chinese first 4th generation (4G) mobile communication trial network for Cooperative Communications was successfully developed as shown in Fig. 1.12, and tested by the Ministry of Science and Technology of the People's Republic of China (MOST). The success of 4G trial network attributes to the National "863" key project—"Future universal technologies for radio environment (FuTURE)".

Furthermore, the trial system throughput was improved up to 1 Gbps in 2009 via new hardware platform and key technologies with enhanced air interface parameters, frame protocol, wireless transmission link and physical layer key technologies of MIMO and LDPC coding. 1 Gbps peak data rate also reached the

Fig. 1.11 Group cell trial systems (multi-antenna units)

highest requirements of ITU IMT-Advanced. Most of the research outcomes are accepted by international and domestic standard organizations by standard proposals and are also verified on the Gbps Trial System.

1.6 Outline

The Group Cell architecture is based on distributed cellular architecture with multi-antenna techniques for cooperative communications, which can enlarge the coverage and employ the advantages of multi-antennas techniques. The Group Cell architecture also has the flat network architecture for IP services. These merits ensure Group Cell implemented in China 4G TDD Trial Systems with physical layer techniques of orthogonal frequency division multiple access (OFDMA) and MIMO etc. The system capacity needs to be analyzed in details and the performances also need to be evaluated.

Moreover, based on this novel cellular architecture, the traditional radio resource management (RRM) strategies need to be evolved too. The handover strategy, power allocation, etc. used in traditional systems cannot accommodate the features of Group Cell architecture, which also need to be improved correspondingly.

The following parts will focus on the construction methods of Group Cell and the RRM strategies.

Chapter 1 covers the definitions, concepts, construction methods and typical application scenarios of Group Cell architecture. The research outcomes of Group Cell architecture for cooperative communications are briefly introduced. The structure of this book is also included in Fig. 1.13. Focusing on the Group Cell architecture, the research outcomes about Capacity Analysis, Slide Handover, Power Allocation strategy of the new cellular structure will be illustrated in details.

Fig. 1.12 Group cell trial systems (two BSs)

The research on CoMP technology inside of Group Cell framework also can be found in each chapter. The Group Cell Trial System supporting cooperative communication will be introduced at the end.

Chapter 2 focuses on the system performance analyses of Group Cell structure through capacity aspects. Since cooperative communication is introduced inside of Group Cell, the system capacity gain also needs to be determined. The Shannon capacity is used to show the system performance comparing with traditional cellular structures. The theoretical analyses are provided to give the capacity gain of

Fig. 1.13 Structure of the
book

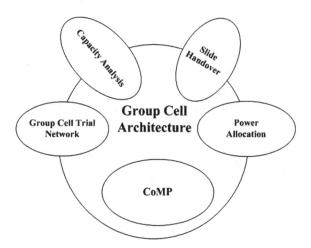

Group Cell. The capacity maximization under individual peak-power constraints for all cooperative transmission nodes in Group Cell architecture is also proposed.

Chapter 3 describes the Slide Handover strategy, which is an important feature of Group Cell architecture to reduce the handover numbers. The reduction of handover can improve the system performance and also the cell edge user performance. The traditional problem about the cell edge effect could be mitigated. The typical application scenarios of Slide Handover will be introduced as urban densely deployed area and highway environments. The system performance with Slide Handover also has improvements with the throughputs and user capacity. Moreover, the simplified version of Slide Handover, fast cell group selection (FCGS) scheme, will be illustrated as the technical solution for 3GPP LTE standards. FCGS was firstly proposed in 2005 and is actually the predecessor of CoMP.

Chapter 4 focuses on power allocation of Group Cell architecture to mitigate the inter-cell interference and maximize system capacity. Power control plays an important role in exploiting the capacity potential. The control of the power transmitted by the BSs has been proposed to reduce the output power level and thus to reduce interference and increase the battery time of the equipment. The dynamic coordination among multiple cells with power limitations of each cell in downlink Group Cell scenario is researched to maximize system capacity.

The Group Cell for Cooperative Communications trial network based on the Group Cell architecture is introduced in Chap. 5. The trial system frame design, physical technology and physical-link design, the performance simulation and evaluation, system parameter design and field-test results will be described. The trial system architecture and the wireless services demonstration results will be presented.

Chapter 2
Capacity Analysis

The capacity analysis is one of the most important methods to demonstrate the system performance. The methods used for analyzing the system capacity include Shannon capacity analysis, outage capacity analysis and cutoff capacity etc. For actual system analysis, the throughput or user number capacity analysis is usually considered. In this chapter, Shannon capacity analysis for Group Cell will be introduced. Multi-user diversity (MUD) can also be used in Group Cell architecture to improve the system capacity. The capacity maximization method with peak-power constraints for cooperative communications will be illustrated.

2.1 Capacity Analysis of Single-User in Group Cell

We will analyze the Shannon capacity of fixed Group Cell structure [1] and slide Group Cell structure in the scenario that each antenna's transmission power is equal. There is still another transmission power allocation method—weighted power transmission scheme [2], which is analyzed in other works. The outage capacity performance analyses will be introduced in Sect. 2.1.2.

2.1.1 Shannon Capacity Analyses of Group Cell

2.1.1.1 Single Group Cell without Shadow Fading

If there is no shadow fading in consideration, the cell selection is only based on the path loss. In a single Group Cell with N antennas, the received power in downlink is $P_r = \sum_{i=1}^{N} P_t L_i$ where P_t is the transmitted power that is equally allocated by each antenna and L_i is the path loss between the antenna and Mobile Terminal (MT).

X. Tao et al., *Group Cell Architecture for Cooperative Communications*, SpringerBriefs in Computer Science, DOI: 10.1007/978-1-4614-4319-3_2, © The Author(s) 2012

If neighboring cell interferences are neglected, the Shannon capacity using equal transmission power scheme can be derived by Shannon formula, which represents an upper bound of the reliable bit transmission rate with limited bandwidth B and the noise power density N_0:

$$C = B\log_2(1 + \sum_{i=1}^{N} P_{t,i}L_i/N_0B) \tag{2.1}$$

The average channel capacity of Group Cell is:

$$C_{AVG} = \iint\limits_{S} p(x,y)B\log_2(1 + \sum_{i=1}^{N} P_tL_i(x,y)/N_0B)dxdy \tag{2.2}$$

where S is the coverage area of the Group Cell and $p(x,y)$ is the probability density function (PDF) of the MT's location in position $z = (x,y)$.

2.1.1.2 Multi-Group Cell with Shadow Fading

In a single Group Cell scenario, the path loss with shadow fading is $L_i' = L_i \times L_{shadow}$, where $L_{shadow} = 10^{-N(0,\sigma^2)/10}$ and σ is the standard deviation of the shadow fading.

Considering Multi-Group Cell scenario, the shadow fading has impacts on the Group Cell selection. For analyzing this effect, a Monte Carlo simulation is used in Multi-Group Cell scenario with center cell consideration only. Equivalent path-loss (EPL) is used as an index of a selected cell.

$$L_{k,EP}' = \sum_{i=1}^{N} L_{k,i}' = \sum_{i=1}^{N} L_{k,i} \times L_{i,shadow} \tag{2.3}$$

where $L_{k,i}'$ is co-operation of path-loss and shadow fading between user and the ith antenna of kth Group Cell.

So, (2.2) can be deduced to:

$$C_{AVG} \approx \frac{\iint_{S,EPL_0 > EPL_k, \forall k \neq 0} B\log_2(1 + P_r/N_0B)\,dxdy}{\left(\frac{9\sqrt{3}}{2}r\right)^2} \tag{2.4}$$

where S is the research area, r is the radius of the Group Cell, EPL_k is the EPL of the kth Group Cell and EPL_0 is the largest of EPL_k.

2.1.1.3 Performance Analysis and Evaluation

The main characteristics for the simulation, used in MATLAB, of the Group Cell architecture and traditional cellular structure are summarized in Table 2.1.

Table 2.1 Simulation parameters of single-user group cell

Parameters	Settings
Group cell size	3
Number of group cells	9
Carrier frequency	5.3 GHz
Total transmission power of AP/BS	43 dBm
Total bandwidth	10 MHz
Bandwidth allocated to each user	100 kHz
Radius of cell	500 m
Thermal noise	−145 dBw
Standard deviation of shadow fading	5 dB

Fig. 2.1 Channel capacity of group cell vs. traditional structure

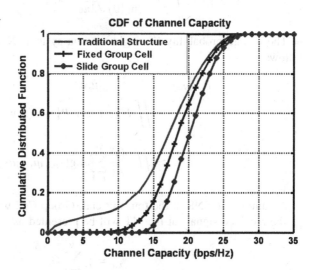

The simulation results are shown in Fig. 2.1, which indicates the cumulative distribution function (CDF) of channel capacity of traditional cellular structure, fixed Group Cell and slide Group Cell structure respectively.

Figure 2.1 shows the gain of channel capacity brought by Group Cell architecture. We can see from the CDF curve that fixed Group Cell can achieve about 50 % gain compared to traditional cellular structure and the slide Group Cell can further achieve 30 % gain vs. fixed Group Cell. The total transmission power of an AP in Group Cell architecture and a BS in tradition structure is equal and the multi-user interferences are ignored in this simulation.

2.1.2 Outage Capacity Analysis of Group Cell

In order to evaluate the coverage performance of Group Cell, we analyze the outage performance [3] of a Group Cell and traditional cellular structure. The outage probability is defined as the probability when received signal to noise ratio

(SNR) is smaller than the target SNR. In this section, we only focus on the analyses of downlink outage performance as an example. Moreover, we assume that the system is free from user interference. As there are more than one antennas that transmit signals to the user in downlink, we normalize the whole power allocated to the user for maintaining the fairness to traditional cellular structure with single antenna.

Based on the power distribution function of Rayleigh-Log-Normal channel model [4]:

$$f_\gamma(\gamma|u,\sigma) = \int_0^\infty \frac{1}{s}e^{-\frac{\gamma}{s}}\frac{10}{\ln(10)\sqrt{2\pi}\sigma s}\exp\{-\frac{(10\log_{10}s - u)^2}{2\sigma^2}\}ds \qquad (2.5)$$

The outage probability of single antenna system ($N = 1$) can be deduced as follows:

$$\begin{aligned} P_{otg,N=1} &= P\{\gamma < \gamma_{req}|\mu,\sigma\} \\ &= \iint_S F_\gamma(\gamma|s,u,\sigma)P\{s = (x,y)|\mu,\sigma\}ds \\ &\approx \iint_S \frac{p(s)}{\sqrt{\pi}}\sum_{i=1}^{N_H}A_i g(x_i,\gamma|\mu,\sigma)ds \end{aligned} \qquad (2.6)$$

Assuming the SNR at the transmit end (Group Cell with N single antennas) is γ_t, the received signals at the MT will be presented as:

$$\gamma_{rec} = \gamma_t \sum_{i=1}^N G_i \qquad (2.7)$$

where G_i is the path gain of the link between the ith antenna and MT. If $\gamma_t = 1$,

$$\begin{aligned} P_{otg} &= \int_{-\infty}^{+\infty} \cdots \int_{-\infty}^{\infty} \frac{1}{\pi^{N/2}}e^{-\sum_{i=1}^N x_i^2} \\ &\quad \{1 - \sum_{i=1}^N e^{-\eta\phi_i}\prod_{n\neq i}\frac{\phi_n}{\phi_n - \phi_n}\}dx_1 dx_2 \cdots dx_N \end{aligned} \qquad (2.8)$$

where $\phi_i = 10^{-(\sqrt{2}\sigma x_i + u_i)/10}$.

Different transmission power allocation methods will have different outage performances. In this section, we only analyzed the equal transmission power allocation scheme like the capacity analysis in Sect. 2.1.1.

With equal transmission power allocation scheme in downlink, each antenna of N total antennas uses equal transmission power to transmit signals. From (2.6) and (2.8), the outage probability of equal transmission power allocation scheme can be expressed as:

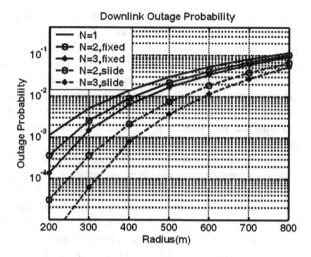

Fig. 2.2 Downlink outage performance with equal transmission power allocation scheme

$$P_{otg} = \int\limits_{-\infty}^{+\infty} \cdots \int\limits_{-\infty}^{\infty} \frac{1}{\pi^{N/2}} e^{-\sum\limits_{i=1}^{N} x_i^2} \bullet$$

$$\{1 - \sum_{i=1}^{N} e^{-\eta\phi_i} \prod_{n \neq i} \frac{10^{-(\sqrt{2}\sigma x_k + u_k)/10}}{10^{-(\sqrt{2}\sigma x_k + u_k)/10} - 10^{-(\sqrt{2}\sigma x_i + u_i)/10}}\} dx_1 dx_2 \cdots dx_N \qquad (2.9)$$

Using multiple Gauss-Hermite integral method, the outage performance of the Group Cell can be derived. We have used Matlab to perform these calculations. The basic parameters of simulation are same as given in Table 2.1. We have analyzed the outage performance of the Group Cell when the size of Group Cell is 1, 2 and 3. The situation of 1 cell in the Group Cell is equal to 1 traditional cellular structure. The simulation results are shown in Fig. 2.2.

It indicates the numerical results of outage performances of fixed and slide Group Cell using equal transmission power allocation scheme in downlink. The figure verifies the merits of Group Cell from the aspect of coverage performance. From the simulation results, we can see that in order to get 1e-2 outage probability, the cell radius of fixed and slide Group Cell of 2 antennas reaches 420 m and 540 m respectively while the cell radius of traditional cellular structure is less than 400 m. When the size of Group Cell reaches to 3, the cell radius of fixed and slide Group Cell structure extends even further.

2.2 Capacity Analysis with Multi-User Diversity in Group Cell

Section 2.1 analyzed the channel capacity and outage performance of the Group Cell architecture. Moreover, MUD is convincingly an effective method to improve system throughput at a large scale. The MUD can be achieved by proper scheduling algorithms in actual mobile communication systems, such as the Maximal C/I algorithm, Proportional Fairness algorithm and so on [5]. But in multi-antenna systems, such as the Group Cell architecture, how to use the MUD is still the emphasis of research activities. There are still several unanswerable questions in this research area, such as the multi-dimensional resource scheduling and allocation, the feedback of user current information and the method of maximizing MUD etc. In this section, we will utilize the MUD strategy in Group Cell and analyze the system performance of the capacity gain brought by MUD strategy.

Consider the scenario of MUD strategy applied in Group Cell architecture. In traditional single antenna cellular environment, by using the strategy of MUD, the capacity of the system will be increased greatly. So, we will analyze the capacity gain in Group Cell architecture with MUD in the following paragraphs. Considering a more universal situation, just like a distributed MIMO [6] case in Group Cell architecture, the result of this analysis will be more useful.

With the strategy of MUD, the system needs to maximize the sum capacity defined as the maximum achievable sum of long-term average data rates transmitted to all users. In the single-input single-output (SISO) case, it has been shown that transmitting to the user with the strongest channel gain in the given time slot is an approach that can achieve the maximum capacity [7]. Therefore for the receivers in SISO system, it is sufficient to feedback their instantaneous SINR to the transmitter. The transmitter then selects the user with the best SINR for transmission.

The challenge of multi-antenna systems is that the link throughput depends on both the received SINR as well as the subspace structure of the channel matrix and the receiver. So, the optimal MUD strategy for allocating antennas and time slots to users is still the hot research area. In this section, we can still use the scheduling algorithms in SISO for multi-user Group Cell architecture. Although this scheduling algorithm for MUD may not be optimal for multi-antenna system, it can still bring the capacity gain at a large scale.

Assume that the transmit end (Group Cell) has Mt antennas and the receive end (MT) has Mr antennas. The system capacity of the distributed MIMO Group Cell architecture can be denoted as:

$$C(Mt, Mr) = \log_2 \det(IMr + \frac{P_r}{N_0 BMt} HH^H) \tag{2.10}$$

where H is a $Mt \times Mr$ complex matrix representing the channel situation between the Mt antennas and Mr antennas. The H^H is the Hermit inverse of H. So, formula will be revised as:

$$C(M_t, M_r)_{AVG} = \iint_S p(x,y) \log_2 \det \left(I_{M_r} + HH^H \frac{\sum_{i=1}^{M_t} P_{t,i} L_i(x,y)}{N_0 BM_t} \right) dxdy \tag{2.11}$$

Moreover, the affection of shadow fading must also be considered because it has some impacts on the power loss and cell selection. Using the customary stationary log-normal model for shadow fading, for all MT position $z = (x, y)$, $L_{shadow} \sim N(0, \sigma^2)$.

The average channel capacity will be:

$$C(M_t, M_r)_{AVG} = \iint_S p(x, y) \log_2 \det \left(I_{M_r} + HH^H \frac{\sum\limits_{i=1}^{M_t} P_{t,i} L_i'(x, y)}{N_0 B M_t} \right) dxdy \quad (2.12)$$

where $L_i'(x, y) = L_i(x, y) \times L_{shadow}$.

Applying the strategy of MUD in the Group Cell, assume that there are K users in a Group Cell. The average channel capacity with MUD will be:

$$C(K, M_t, M_r)_{AVG} = \max_{k=1,2,\ldots,K} \left[\iint_S p(x, y) \log_2 \det \left(I_{M_r} + H_k H_k^H \frac{\sum\limits_{i=1}^{Mt} P_{t,i} L_i'(x, y)}{N_0 B M_t} \right) dxdy \right]$$

$$(2.13)$$

2.2.1 Calculation of Capacity Gain with MUD

The use of MUD strategy will greatly improve the capacity of Group Cells. Let's analyze the capacity gain brought by MUD in detail.

At first, let's assume a simpler model of MIMO in which number of antennas at the transmitter is equal to the number of antennas at the receiver, which means $Mt = Mr = M$.

So, formula (2.10) will be revised as:

$$C(M) = \sum_{j=1}^{M} \log_2 \left(1 + \frac{P_r \lambda_j}{N_0 B M} \right) \quad (2.14)$$

where λ_j is the jth eigenvalue of the matrix $W = HH^H$. So, formula (2.13) will be revised as:

$$C(K, M)_{AVG} = \max_{k=1,2,\ldots,K} \left[\iint_S p(x, y) \sum_{j=1}^{M} \log_2 \left(1 + \frac{\lambda_j \sum\limits_{i=1}^{M} P_{k,t,i} L_{k,i}'(x, y)}{N_0 B M} \right) dxdy \right]$$

$$(2.15)$$

From this formula, the exact capacity gain cannot be calculated in a simple form. But we can analyze its upper bound and lower bound.

Assuming that $\gamma k(x,y) = \sum_{i=1}^{M} P_{k,t,i} L'_{k,i}(x,y) \Big/ N_0 BM$ represents the SNR of the kth user, the formula will be simpler. We can easily get the upper-bound and lower-bound of the average capacity from the formula mentioned above.

$$C(K,M)_{AVG} \leq M \max_{k=1,2,...,K} \left(\iint_S p(x,y) \log_2(1 + \lambda_{k,\max} \gamma_k(x,y)) dx dy \right)$$

and

$$C(K,M)_{AVG} \geq M \max_{k=1,2,...,K} \left(\iint_S p(x,y) \log_2(1 + \lambda_{k,\min} \gamma_k(x,y)) dx dy \right) \quad (2.16)$$

where $\lambda_{k,\min}$ and $\lambda_{k,\max}$ are the minimum and maximum eigenvalues of the matrix $W = HkHk^H$ respectively.

Further calculations will require the PDF of $\lambda_{k,\min}$ and $\lambda_{k,\max}$. The PDF of $\lambda_{k,\min}$ can be taken from [8] and it is an exponential distribution with M being its mean. The lower bound of $C(K,M)_{AVG}$ can be deduced further, but the PDF of $\lambda_{k,\max}$ has no determinate form. So, we can use the numerical analysis to obtain the upper bound of $C(K,M)_{AVG}$ in order to get the approximate result of the capacity gain, brought by MUD.

For the lower bound of $C(K,M)_{AVG}$, let's analyze two situations. When $\gamma_k(x,y) \cdot 1$, formula can be revised as:

$$C(K,M)_{AVG} \geq M \max_{k=1,2,...,K} \left(\iint_S p(x,y) \log_2(\lambda_{k,\min} \gamma_k(x,y)) dx dy \right) \quad (2.17)$$

Lemma 1 Consider independently and identically distribute (i.i.d.) random variables $\xi 1, \xi 2, ... \xi K$ with CDF $F(\cdot)$ and PDF $f(\cdot)$.

If $\lim_{x \to \infty} \{[1 - F(x)]/f(x)\} = c > 0$, where c is a constant.

Then,

$$\max_{k=1,2,...,K}(\xi_1, \xi_2, ... \xi_K) - l_K \xrightarrow{d} \exp[-\exp(-x/c)]$$

with $F(lK) = \int_{-\infty}^{lK} f(x) dx = 1 - 1/K$.

Furthermore, the value of l_K can be derived from (2.7) that $l_K = \ln K$.

From Lemma 1 and the PDF of $\lambda_{k,\min}$, $\max_{k=1,2,...,K}[\lambda_{k,\min}] - \ln K$ leads to a variable whose PDF is $\exp[-\exp(-x/c)]$. So, the formula (2.17) can be revised as:

$$C(K,M)_{AVG} \geq M\left(\ln K + \iint_S p(x,y)\log_2 \gamma k(x,y)dxdy\right) \qquad (2.18)$$

For a comparison with the Group Cell without MUD, the lower bound of a capacity gain of each antenna dimension can be written as:

$$Gain_{min} = \frac{C(K,M)_{AVG}}{M} - C_{AVG}$$

$$\geq \left(\ln K + \iint_S p(x,y)\log_2 \gamma k(x,y)dxdy\right)$$

$$- \iint_S p(x,y)\log_2(1 + \gamma k(x,y))dxdy \qquad (2.19)$$

$$\approx \ln K$$

When $\gamma k(x,y) \to 0$, formula (2.19) can be revised as:

$$C(K,M)_{AVG} \geq M \iint_S p(x,y)\gamma k(x,y) \max_{k=1,2,...,K}(\lambda_{k,\min})dxdy \qquad (2.20)$$

From Lemma 1 and the PDF of $\lambda_{k,\min}$ formula (2.20) can be simplified further.

$$C(K,M)_{AVG} \geq M \ln K \iint_S p(x,y)\gamma k(x,y)dxdy \qquad (2.21)$$

For a comparison with the Group Cell without MUD, the lower bound of a capacity gain of each antenna dimension is

$$Gain_{min} = \frac{C(K,M)_{AVG}/M}{C_{AVG}} \geq \frac{\ln K \int_S p(x,y)\gamma k(x,y)dxdy}{\int_S p(x,y)\log_2(1 + \gamma k(x,y))dxdy} \qquad (2.22)$$

$$\approx \ln K$$

In the following part, we will use the simulation approach to verify the theoretical analysis stated above.

Table 2.2 Simulation
parameters of MUD in group
cell

Parameters	Settings
Group cell size	3
Number of group cells	9
Carrier frequency	5.3 GHz
Total transmission power of AP/BS	20 W
Total bandwidth	10 MHz
Bandwidth allocated for each user	100 kHz
Radius of cell	500 m
Thermal noise	−145dBw
Standard deviation of shadow fading	5 dB

2.2.2 Performance Analysis and Evaluation

The main characteristics of the simulation for the Generalized Distributed Cellular
Architecture—Group Cell are summarized in Table 2.2.

In the simulation, the MT and Group Cell are all deployed with single antenna.
The simulation results are shown in Figs. 2.3 and 2.4, which indicates the performance gain of Group Cell vs. traditional cellular structure and the capacity gain
brought by MUD, respectively.

Figure 2.3 gives the capacity of Group Cell with MUD. From the simulation
results, we can see that the Group Cell capacity with MUD is improved as the
number of users increased.

To lend further credence to capacity gain brought by MUD, we plot in Fig. 2.4
with the comparison to the bound $\ln K$. From the simulation results, with the
increase of user numbers, the capacity gain brought by MUD follows the bound
very well. The slight decrease of MUD gain to $\ln K$ is due to the simulation
settings that all single antennas deployed in MT and AP.

2.3 Ergodic Capacity of Group Cell Systems with Power Constraints

The system capacity is a very important requirement from perspective of the
system performance viewpoint. The target peak data rate in the downlink is 1 Gb/s
for LTE-Advanced and the target peak data rate in the uplink is 500 Mb/s considering the traffic demands in cellular networks. Capacity analysis gives us the
direction of how to maximize system capacity according to different situations and
based on different criterions. In physical implementations of Group Cell systems,
each eNodeB has its own power amplifier in its analog front-end and is limited
individually by the linearity of the power amplifier. For example, in the standardization of 3GPP LTE-Advanced, it is emphasized that each eNodeB has its
own maximal transmit power of 49 dBm for 40 MHz LTE-A carrier [9].

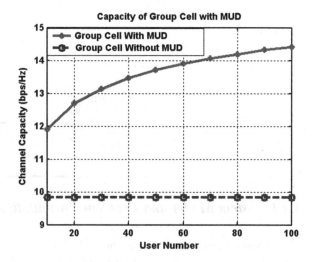

Fig. 2.3 Capacity of group cell with MUD

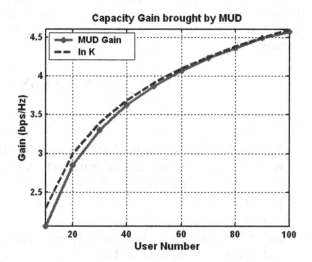

Fig. 2.4 Capacity gain brought by MUD

Furthermore, it is also pointed out that individual peak-power constraints (IPC) must be applied to a group of antennas on each eNodeB [10]. Hence, an IPC on a per-transmit- eNodeB basis is needed for the practical implementation:

$$E(|\mathbf{X}_i|^2) \leq P_i \quad for \ i = 1, 2, \ldots, N. \qquad (2.23)$$

where $\mathbf{X}_i$ denotes the transmitted signal from the ith BS, $E(\cdot)$ means the expectation operator, and P_i is the ith eNodeB's peak-power constraint.

Fig. 2.5 Group cell system model with (N, L_i, M)

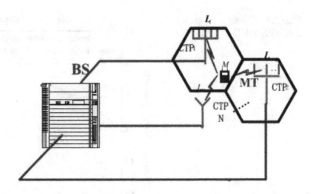

2.3.1 System Model and Problem Formulation

The Group Cell system model is shown in Fig. 2.5, which consists of N largely separated coordinated transmission points (CTPs) each with L_i ($i = 1, 2,..., N$) centralized antennas and one user equipped with M antennas. The channel information is known at transmitter and receiver. Assume each signal transmitted from every CTP arrive at the user simultaneously, the downlink received signal is expressed as

$$\mathbf{Y} = \sum_{i=1}^{N} \sqrt{\eta_i} \mathbf{H}_i \mathbf{X}_i + \mathbf{n} = \mathbf{H}\mathbf{X} + \mathbf{n} \tag{2.24}$$

where

1. $\mathbf{X} = (\mathbf{X}_1, \mathbf{X}_2, \ldots, \mathbf{X}_N)^T$ are transmitted signals from N CTPs.
2. $\mathbf{H}_i \in \mathbb{C}^{M \times L_i}$ indicates the small-scale fading from ith CTP to the user, assumed to be frequency-flat fading with zero mean and unit variance; $\mathbf{H} = \left(\sqrt{\eta_1} \mathbf{H}_1 \quad \sqrt{\eta_2} \mathbf{H}_2 \quad \cdots \quad \sqrt{\eta_N} \mathbf{H}_N \right)$ represents a $M \times \sum_{i=1}^{N} L_i$ compound channel;
3. $\mathbf{n} \in \mathbb{C}^{M \times 1}$ is an independent circularly symmetric complex Gaussian noise vector with distribution $\mathcal{CN}(0, \mathbf{I}_M)$;
4. η_i are the parameters related to the SNR

$$\eta_i = \frac{SNR_i}{P_i} \tag{2.25}$$

Where SNR_i is the normalized power ratio of $\mathbf{X}_i$ to the noise (after fading); P_i denotes the individual peak-power constraint of the ith CTP.

This subsection targets the capacity-maximization problem under a more realistic condition, i.e. IPC, in which each BS has its own transmit peak-power constraint. According to (2.24), the instantaneous capacity can be expressed as

$$C = \log\left|\mathbf{I}_M + \mathbf{H}\mathbf{Q}\mathbf{H}^\dagger\right| \overset{(a)}{=} \log\left|\mathbf{I} + \mathbf{Q}\mathbf{H}^\dagger\mathbf{H}\right| \tag{2.26}$$

where

1. (a) comes from the identity $|\mathbf{I} + \mathbf{AB}| = |\mathbf{I} + \mathbf{BA}|$;
2. $\mathbf{Q}$ is the covariance expression of transmitted signals:

$$\mathbf{Q} = E(\mathbf{X}\mathbf{X}^\dagger) = \begin{pmatrix} \mathbf{Q}_{11} & \mathbf{Q}_{12} & \cdots & \mathbf{Q}_{1N} \\ \mathbf{Q}_{21} & \mathbf{Q}_{22} & \cdots & \mathbf{Q}_{2N} \\ \cdots & \cdots & \cdots & \cdots \\ \mathbf{Q}_{N1} & \mathbf{Q}_{N2} & \cdots & \mathbf{Q}_{NN} \end{pmatrix} \tag{2.27}$$

where $\mathbf{Q}_{ij}$ denotes the covariance matrix between signals from the ith and jth CTPs. We consider IPC for each CTP, i.e. $tr(\mathbf{Q}_{ii}) \leq P_i$. To maximize the capacity in (2.25), the solution is to optimize the transmit covariance matrix $\mathbf{Q}$, meanwhile the IPC condition is satisfied:

$$\{\mathbf{Q}^*\} = \arg\max C$$
$$s.t.\ tr(\mathbf{Q}_{ii}) \leq P_i\ for\ i = 1, 2, ...N \tag{2.28}$$

2.3.2 Capacity Analysis for Group Cell with Power Constraints

2.3.2.1 Capacity Analysis for (N, L_i, M) Group Cell systems

Since $\mathbf{H}^\dagger\mathbf{H} \in \mathbb{C}^{\sum_{i=1}^N L_i \times \sum_{i=1}^N L_i}$ is a positive semi-definite Hermitian matrix, we can diagonalize it and write $\mathbf{H}^\dagger\mathbf{H} = \mathbf{U}\mathbf{D}\mathbf{U}^\dagger$ in which $\mathbf{D} \in \mathbb{R}^{\sum_{i=1}^N L_i \times \sum_{i=1}^N L_i}$ is diagonal matrix with descending nonnegative values and $\mathbf{U} \in \mathbb{C}^{\sum_{i=1}^N L_i \times \sum_{i=1}^N L_i}$ is unitary. Thus (2.26) is transformed into

$$C = \log\left|\mathbf{I} + \mathbf{Q}\mathbf{U}\mathbf{D}\mathbf{U}^\dagger\right| = \log\left|\mathbf{I} + \mathbf{U}^\dagger\mathbf{Q}\mathbf{U}\mathbf{D}\right| \tag{2.29}$$

Define $\mathbf{S} \triangleq \mathbf{U}^\dagger\mathbf{Q}\mathbf{U}$, then $\mathbf{Q} = \mathbf{U}\mathbf{S}\mathbf{U}^\dagger$, hence (2.29) becomes $C = \log|\mathbf{I} + \mathbf{S}\mathbf{D}|$. Using the Hardamard inequality [11],

$$|\mathbf{I} + \mathbf{S}\mathbf{D}| \leq \prod_{k=1}^{\sum_{i=1}^N L_i} (1 + S_{kk}D_{kk}) \overset{(a)}{=} \prod_{k=1}^{K} (1 + S_{kk}D_{kk}) \tag{2.30}$$

(a) owes to the fact that $D_{kk} = 0$ for $k > K$, $rank(\mathbf{H}) = K$. D_{ij} means the element of ith column and jth row in matrix $\mathbf{D}$ and the same is with S_{ij}. As verified in [12], in order to find all solutions to maximize the instantaneous capacity in (2.25), it is sufficient to consider only the class of diagonal, positive semi-definite matrices $\mathbf{S}$ that satisfies $S_{ij} = 0$ for all $i, j > K$. In other words, the solution is to find the matrix $\mathbf{S}$ with the form $\mathbf{S} = \begin{pmatrix} \mathbf{S}_1 & \mathbf{0} \\ \mathbf{0} & \mathbf{0} \end{pmatrix}$, where $\mathbf{S}_1 = diag(S_{kk})$, $k = 1, 2, \ldots, K$. Hence, the capacity-maximization problem is transformed into the following

$$\max_{\{S_{kk}\}_{k=1}^{K} : S_{kk} \geq 0} \sum_{k=1}^{K} \log(1 + S_{kk} D_{kk}) \tag{2.31}$$

Next, we will derive the equivalent transmit peak-power constraints. Divide $\in \pounds^{\sum_{i=1}^{N} L_i \times \sum_{i=1}^{N} L_i}$ into N parts, i.e., let the first L_1 rows of $\mathbf{U}$ to be $\mathbf{u}_1$ and so on. From the expression of $\mathbf{Q} = \mathbf{USU}^\dagger$, $\mathbf{Q}$ will be converted into

$$\begin{pmatrix} \mathbf{Q}_{11} & \mathbf{Q}_{12} & \cdots & \mathbf{Q}_{1N} \\ \mathbf{Q}_{21} & \mathbf{Q}_{22} & \cdots & \mathbf{Q}_{2N} \\ \cdots & \cdots & \cdots & \cdots \\ \mathbf{Q}_{N1} & \mathbf{Q}_{N2} & \cdots & \mathbf{Q}_{NN} \end{pmatrix} = \begin{pmatrix} \mathbf{u}_1 \\ \mathbf{u}_2 \\ \cdots \\ \mathbf{u}_N \end{pmatrix} \mathbf{S} \begin{pmatrix} \mathbf{u}_1^\dagger & \mathbf{u}_2^\dagger & \cdots & \mathbf{u}_N^\dagger \end{pmatrix} \tag{2.32}$$

where $\mathbf{u}_i$ is the ith sub-matrix of $\mathbf{U}$ with dimension of L_i rows, hence the IPC condition can be rewritten

$$tr(\mathbf{Q}_{ii}) = tr(\mathbf{u}_i \mathbf{S} \mathbf{u}_i^\dagger) = \sum_{k=1}^{K} c_{i,k} S_{kk} \leq P_i \tag{2.33}$$

where $c_{i,k}$ is the non-zero eigenvalue of $\mathbf{u}_i(:, k) \mathbf{u}_i(:, k)^\dagger$. $\mathbf{u}_i(:, k)$ means the kth column of $\mathbf{u}_i$ $i = 1, 2, \ldots, N$. Therefore the (N, L_i, M) system capacity maximization problem is

$$\max_{\{S_{kk}\}_{k=1}^{K} : S_{kk} \geq 0} \sum_{k=1}^{K} \log(1 + S_{kk} D_{kk})$$
$$s.t. \ tr(\mathbf{u}_i \mathbf{S} \mathbf{u}_i^\dagger) = \sum_{k=1}^{K} c_{i,k} S_{kk} \leq P_i, \ i = 1, 2, ..., N \tag{2.34}$$

Obviously, the optimization problem is concave in $\mathbf{S}$, and the constraint condition is convex, thus (2.34) can be solved by convex programming [13].

2.3.2.2 Capacity Analysis for $(2, L_i, M)$ Group Cell Systems

To better elaborate the system's capacity, we derive the explicit solution to maximize capacity for the special case where two BSs cooperate to transmit signals to the user. According to (2.33), the $(2, L_i, M)$ Group Cell system maximal capacity problem can be written as:

$$f = \min_{\{S_{kk}\}_{k=1}^{K}:S_{kk}\geq 0} -\sum_{k=1}^{K}\log(1 + S_{kk}D_{kk})$$

$$s.t. \sum_{k=1}^{K}c_{1,k}S_{kk} \leq P_1, \sum_{k=1}^{K}c_{2,k}S_{kk} \leq P_2 \tag{2.35}$$

It can be solved via Lagrangian function, the Lagrangian function is

$$g(\lambda, \mu) = -\sum_{k=1}^{K}\log(1 + S_{kk}D_{kk}) - \sum_{k=1}^{K}\lambda_k S_{kk}$$

$$+ \mu_1(\sum_{k=1}^{K}c_{1,k}S_{kk} - P_1) + \mu_2(\sum_{k=1}^{K}c_{2,k}S_{kk} - P_2) \tag{2.36}$$

where both $\lambda = (\lambda_1, \lambda_2, \ldots, \lambda_K)^T$ and $\mu = (\mu_1, \mu_2)^T$ are Lagrange multiplier column vectors associated with the inequality constraints.

The KKT condition is as follows:

$$\begin{cases} -\frac{D_{kk}}{1+S_{kk}D_{kk}} - \lambda_k + \mu_1 c_{1,k} + \mu_2 c_{2,k} = 0 & (a) \\[2mm] \lambda_k S_{kk} = 0 & (b) \\[2mm] \mu_1(\sum_{k=1}^{K}c_{1,k}S_{kk} - P_1) = 0, \mu_2(\sum_{k=1}^{K}c_{2,k}S_{kk} - P_2) = 0 & (c) \\[2mm] S_{kk} \geq 0, \lambda_k \geq 0, \mu_1, \mu_2 \geq 0 \ , \ k = 1, 2, \ldots, K & (d) \end{cases} \tag{2.37}$$

We aim to find the solution of $\{S_{kk}\}_{k=1}^{K}$, λ and μ via Eq. (2.37). Assume there are K_1 positive values in $\{S_{kk}\}_{k=1}^{K}$. We can define the index set $I = \{k|S_{kk} \neq 0\}$, the size of I is K_1 and the size of $\bar{I}$ is K_2, $K_1 + K_2 = K$. From the equation (a) in (2.37) we get:

$$\mu_1 c_{1,k} + \mu_2 c_{2,k} = \lambda_k + \frac{D_{kk}}{1 + S_{kk}D_{kk}} \overset{(*)}{\geq} 0 \tag{2.38}$$

(*) owes to (d) in (2.37), hence $\mu = (\mu_1, \mu_2)^T$ is not zero, i.e., at least one of μ_1 and μ_2 is positive. Therefore there are two cases for obtaining the solution:

A. *Case 1: only one of μ_1 and μ_2 is positive.*

Without loss of generality, assume $\mu_1 = 0$, $\mu_2 \neq 0$
   Via derivation of (2.38), we get

$$\mu_2 = \frac{K_1}{P_2 + \sum_{k=1}^{K_1} \frac{c_{2,k}}{D_{kk}}} \tag{2.39}$$

Thus:

(a) for $k \in I$, $S_{kk} = \frac{1}{\mu_2 c_{2,k}} - \frac{1}{D_{kk}}$ and $\lambda_k = 0$.
(b) for $k \notin I$, $S_{kk} = 0$. According to (2.39), there is $\lambda_k = \mu_2 c_{2,k} - D_{kk}$.

   Then when $\mu_1 = 0, \mu_2 \neq 0$, the solution $(\mathbf{S}, \boldsymbol{\lambda}, \boldsymbol{\mu})$ is:

$$\mathbf{S}^*_{Case1} = \begin{cases} S_{kk} = \frac{1}{\mu_2 c_{2,k}} - \frac{1}{D_{kk}}, & k \in I \\ S_{kk} = 0, & k \notin I \end{cases}$$

$$\boldsymbol{\lambda}^*_{Case1} = \begin{cases} \lambda_k = 0, & k \in I \\ \lambda_k = \mu_2 c_{2,k} - D_{kk}, & k \notin I \end{cases} \tag{2.40}$$

$$\mu^*_{1,Case1} = 0, \quad \mu^*_{2,Case1} = \frac{K_1}{P_2 + \sum_{k=1}^{K_1} \frac{c_{2,k}}{D_{kk}}}$$

The value of K_1 is determined via iterative computation until (d) in (2.37) is satisfied. Via substituting the value of (2.40) in (2.39), we obtain the optimum function value in (2.35):

$$f^*_{opt1} = \sum_{k \in I} \log\left(\frac{D_{kk}\left(P_2 + \sum_{k=1}^{K_1} \frac{c_{2,k}}{D_{kk}}\right)}{c_{2,k} K_1}\right).$$

B. *Case 2: none of μ_1 and μ_2 is zero.*

   According to (c) in (2.37), we get:

$$\sum_{k=1}^{K} c_{1,k} S_{kk} = P_1, \quad \sum_{k=1}^{K} c_{2,k} S_{kk} = P_2 \tag{2.41}$$

which means both of the two CTPs transmit signals with full power constraint. Via mathematical derivation o (2.40), we get:

(a) for $k \in I$, $\lambda_k = 0$ and $\mu_1 c_{1,k} + \mu_2 c_{2,k} = \frac{D_{kk}}{1+S_{kk}D_{kk}}$, thus $S_{kk} = \frac{1}{\mu_1 c_{1,k} + \mu_2 c_{2,k}} - \frac{1}{D_{kk}}$;
(b) for $k \notin I$, $S_{kk} = 0$ and $\lambda_k = \mu_1 c_{1,k} + \mu_2 c_{2,k} - D_{kk}$.

   So when none of μ_1 and μ_2 is zero, the solution $(\mathbf{S}, \boldsymbol{\lambda}, \boldsymbol{\mu})$ will be

$$\mathbf{S}^*_{Case2} = \begin{cases} S_{kk} = \frac{1}{\mu_1 c_{1,k} + \mu_2 c_{2,k}} - \frac{1}{D_{kk}}, & k \in I \\ S_{kk} = 0, & k \notin I \end{cases}$$

$$\lambda^*_{Case2} = \begin{cases} \lambda_k = 0, & k \in I \\ \lambda_k = \mu_1 c_{1,k} + \mu_2 c_{2,k} - D_{kk}, & k \notin I \end{cases}$$

(2.42)

The value of $(\mu^*_{1,Case2}, \mu^*_{2,Case2})$ is chosen to satisfy the peak-power constraints:

$$\begin{cases} \sum_{k=1}^{K} c_{1,k} S_{kk} = \sum_{k \in I} (\frac{c_{1,k}}{\mu_1 c_{1,k} + \mu_2 c_{2,k}} - \frac{c_{1,k}}{D_{kk}}) = P_1 \\ \sum_{k=1}^{K} c_{2,k} S_{kk} = \sum_{k \in I} (\frac{c_{2,k}}{\mu_1 c_{1,k} + \mu_2 c_{2,k}} - \frac{c_{2,k}}{D_{kk}}) = P_2 \end{cases}$$

(2.43)

The optimum function value in (2.35) is

$$f^*_{opt2} = \sum_{k \in I} \log(\frac{D_{kk}}{\mu_1 c_{1,k} + \mu_2 c_{2,k}})$$

Conclusion: The maximal capacity of $(2, L_i, M)$ system is $C_{\max} = -\min (f^*_{opt1}, f^*_{opt2})$.

2.3.3 Performance Analysis and Evaluation

Monte Carlo simulations are performed to evaluate the ergodic capacity of Group Cell systems under IPC condition with frequency-flat Rayleigh fading. The mean SNR parameters (namely η_i) play a key role in the simulation.

In the simulations, the user and each BS are equipped with the same number of antennas with $L_i = M = 4 (i = 1, 2, \ldots, N)$; the total power constraint is $P_{total} = 20W$ and each CTP's transmit peak-power is equally confined to $P_i \leq P_{total}/N (i = 1, 2, \ldots, N)$.

Figure 2.6 shows the ergodic capacity verse SNR1 for the link of BS1 to the user for different numbers of cooperative BSs. The SNR for the link of BS2 to the user is SNR2. Also the transmit power control (TPC) based water-filling algorithm in [14] is simulated under the same conditions. As shown in Fig. 2.6, the ergodic capacity in Group Cell systems increases with the number of cooperative BSs. The diversity gain due to cooperation among CTPs easily accounts for the result.

Next, Figs. 2.7 and 2.8 illustrate the ergodic capacity for the special case $N = 2$. For comparison, the TPC-based water-filling algorithm and uniform power allocation scheme are also simulated under the same conditions. Figure 2.7 shows that our proposed IPC-based method outperforms the TPC-based uniform power allocation scheme by about 39% when SNR1 = 0 dB, $\eta_2 = \eta_1$.

Fig. 2.6 Ergodic capacity under $L_i = M = 4, \eta_1 = \eta_i$

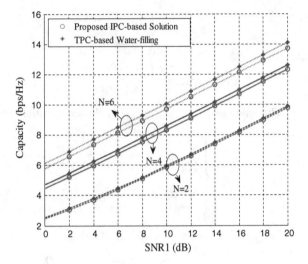

Fig. 2.7 Ergodic capacity under $L_i = M = 4$, $N = 2$, $\eta_1 = \eta_2$

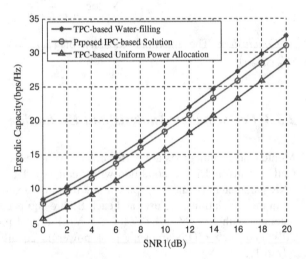

This is because our scheme makes the best use of channel knowledge to realize water-filling under IPC so as to better combat with channel fading. Compared to the TPC-based water-filling, our solution suffers from a little performance loss. The reason is that, the TPC-based water-filling assumes ideal power usage and perfect power cooperation among CTPs. But our IPC-based solution has no power cooperation among BSs, resulting in a less efficient usage of power. However, as each BS must have its own power amplifier in its analog front-end, the IPC is a more realistic condition for practical Group Cell systems. Thereby despite a little performance loss for our solution compared to the TPC-based water-filling, our scheme exhibits more suitability and reliability for Group Cell systems and has more practical implications.

Fig. 2.8 CDF of the capacity under $L_i = M = 4$, $N = 2, \eta_1 = \eta_2$

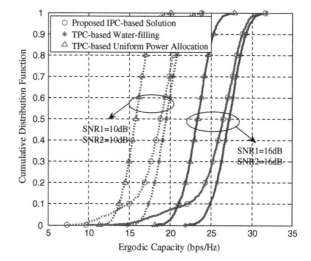

Figure 2.8 shows the CDF of the ergodic capacity under IPC. It is clear that the capacity offered by our IPC-based solution is better than the TPC-based uniform power allocation, and approaches to the TPC-based water-filling. Similar results are seen in Fig. 2.7 which verifies the simulation analysis from Fig. 2.7.

2.4 Summary

The analyses in Sect. 2.1 showed that the Group Cell architecture can get more superiority than traditional cellular structure in terms of system capacity and coverage. The theoretical analyses in Sect. 2.2 are provided to give the capacity gain bought by MUD and the bottom bound of capacity gain is deduced. In Sect. 2.3, we focused on capacity maximization in Group Cell architecture under individual peak-power constraints for all cooperative transmission nodes. Moreover, our result is achieved under IPC, which is more realistic in practical Group Cell implementations. Simulation results showed that our proposed solution outperforms TPC-based uniform power allocation scheme in terms of the ergodic capacity. Also an elegant tradeoff between system capacity and reality significance can be obtained from our proposed capacity-maximization solution.

Chapter 3
Slide Handover

Slide Handover is an important feature of Group Cell architecture. In this chapter, we will introduce the handover schemes that will be suitable for the new cellular architecture—Group Cell, which is Slide Handover mode [31] and Fast Cell Group Selection Scheme [32].

3.1 Slide Handover Mode

3.1.1 Group Cell and Slide Handover

Group Cell, which is introduced in the Chap. 1, is characterized by several adjacent cells which use the same resource to communicate with a certain user and use different resource to communicate with different users. These cells connected to a BS/AP can form one or several Group Cells.

The structure of one Group Cell may be varied in different environments. Figure 3.1 shows the Group Cell structure in highway. This scenario includes more than one Group Cells.

The signal could be transmitted and received by all the antennas of one Group Cell using the techniques that may be applied in the future mobile system such as JT, Space–Time Code (STC), MIMO, OFDMA and DAS. Therefore, the system's ability to resist interference can be improved, the number of handover can be greatly decreased and the system capacity can also be increased significantly. Based on the Group Cell structure, the users do not need a handover when they cross cells lay in one Group Cell. It will not perform the traditional handover, until it moves out of the coverage of one Group Cell where the Group Handover will be performed. Group Handover is defined as the handover from one Group Cell to another Group Cell.

X. Tao et al., *Group Cell Architecture for Cooperative Communications*,
SpringerBriefs in Computer Science, DOI: 10.1007/978-1-4614-4319-3_3,
© The Author(s) 2012

Fig. 3.1 Group cell in highway

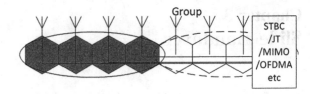

Fig. 3.2 Slide handover process in highway scenario

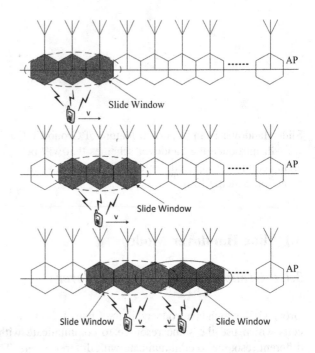

The construction of Group Cell can be flexible according to the definition of a Group Cell given above. In fact, it may be seen as a process of sliding windows. The several cells in one Group Cell can be regarded as in one window and the window can change dynamically in size, shape and sliding speed due to the vehicle speed and direction of the mobile user. When the mobile user moves at rapid speed, the sliding window size will increase to keep up with the motion of the mobile user and decrease the number of handover. When the speed of mobile user is relatively slow, the size of the slide window will become smaller to reduce the wastage of resources. If a mobile user changes its direction, the direction of slide window would be changed at the same time. Figure 3.2 shows the process of sliding windows. As a new handover mode, Slide Handover changes adaptively in correspondence with a Group Cell structure. Different mobile users may correspond to different Group Cells. The construction of one Group Cell is dynamically changed instead of fixed.

Table 3.1 Simulation parameters for slide handover

Parameters	Settings
Service type	Voice
Average call length	90 s
Frame duration	10 ms
Number of time slots	15 slots/frame
Information bit rate	12.2 kbit/s
Bandwidth	5 MHz
Carrier frequency	5G
Spreading factor	128
Distance between two antenna	1 km
Target SINR (uplink)	−19.23 dB
Target SINR (downlink)	−20.73 dB
TX power control step size	1 dB
Thermal noise	−133 dBW

Fig. 3.3 Equivalent pathloss of slide group cell

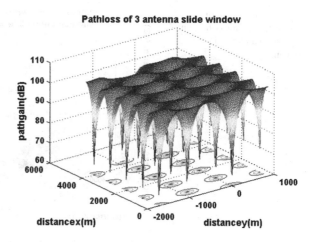

3.1.2 Performance Analysis and Evaluation

Let's assume that the traffic for simulation have an exponential distribution with average duration of each call. The call arrival rate describes the average number of new calls arriving as per antenna per second. The relative traffic load $\rho = \lambda/\mu$, describes how many new calls can be expected to arrive during an ongoing call. The quantity ρ is usually measured in Erlang. Simulation results are obtained in enhanced RUNE simulator in consideration of pathloss and Rayleigh fading. (Table 3.1)

According to our current simulation results, Slide Handover could decrease the equivalent pathloss in downlink (ref. Figs. 3.3, 3.4, 3.5 and 3.6) and noise rise in uplink (ref. Fig. 3.7) so as to improve the system capacity by 10–30 %. (ref. Fig. 3.7). The equivalent pathloss is defined as follows:

Fig. 3.4 Equivalent pathloss of fixed group cell

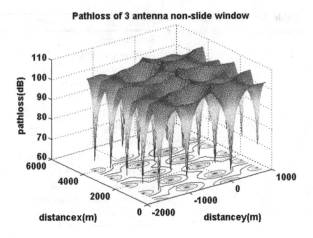

Fig. 3.5 Decrease of equivalent pathloss brought by slide handover in downlink

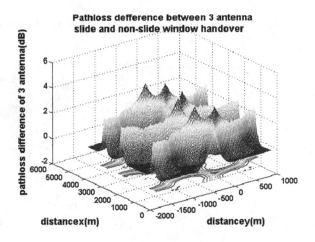

Fig. 3.6 Pathloss difference in group cell architecture

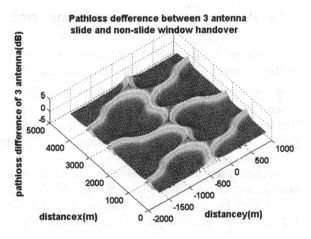

Fig. 3.7 Decrease of noise rise brought by slide handover in uplink

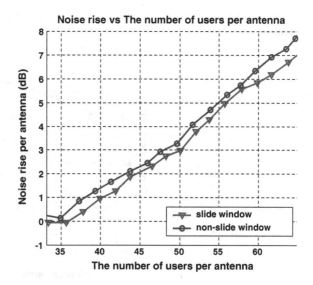

$$Equivalent\ Pathloss = \frac{\sum\limits_i P_{ti}}{P_r} \qquad (3.1)$$

where P_{ti} denotes the transmitted power from the ith antenna element of one Group Cell to a certain user and P_r denotes the user's received power.

To lend further credence to the approaches for Slide Handover in Group Cell infrastructure, we plot Fig. 3.5. The decrease of equivalent pathloss is brought by Slide Handover in downlink in Group Cell infrastructure, which is more direct measure of system capacity in CDMA system level simulations. The results show that maximum pathloss difference is 4.25 dB.

Figure 3.6 uses the same platform of Fig. 3.5 from which we can get the profile of Group Cell with window size 3. The noise rise is defined as the ratio of the total received wideband power to the noise power [33], (Fig. 3.8)

$$Noiserise = \frac{I_{total}}{P_N} \qquad (3.2)$$

3.2 Fast Cell Group Selection Scheme Mode

3.2.1 System Scheme

The Group Cell architecture will be used in the future mobile communication systems. But for the current 3G and Enhanced 3G systems, such as LTE, there still remains the traditional cellular structure [34]. The Group Cell and the Slide Handover cannot fully explore the advantages in current system.

Fig. 3.8 Capacity per antenna

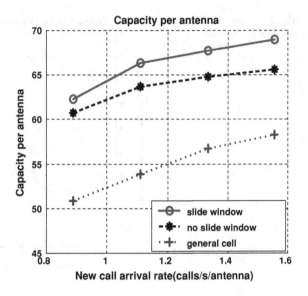

In the process of developing LTE standard, the improvement of cell edge user performance has been the focus of both standardizing work and academic researches. There are already some possible solutions such as Macro Diversity, Fast Cell Selection (FCS) and Interference Mitigation schemes. In order to solve cell edge user performance problem, we proposed a new handover scheme called Fast Cell Group Selection (FCGS) based on the Group Cell and Slide Handover mode in the Aug. of 2005 [31]. Furthermore, the standard proposal based on FCGS for LTE has been submitted to 3GPP TSG (Technical Standard Group) RAN1 (Radio Area Network 1). The cellular structure for FCGS can be current 3G and E3G network architectures and the physical technologies also can be CDMA or OFDM/MIMO, etc.

According to FCGS scheme, the MT measures instantaneous downlink pilot signal power of serving or neighboring cells. When it locates at the cell edge area, it will choose a group of best cells (as the serving Group Cell in the Group Cell architecture) and indicates them to transmit downlink signals by uplink channel. The chosen cell group transmits downlink signals simultaneously and the MT combines its received signals. Since the cell selection and indicating mechanism is the same as FCS, the cell group selection and updating period of FCGS scheme can be as fast as the FCS scheme.

The main point of FCGS is to take full advantages of Macro Diversity and FCS. The chosen group of cells can flexibly transmit the same signals or different signals by advanced signal processing techniques, e.g. Space Time Block Code (STBC) etc. [35, 36]. It will provide Macro Diversity gain to FCS scheme by selecting more than one transmitting cell. Since the cell selection procedure is handled mainly by physical layer with the MT, the frequency of selection and updating can be much faster than that of Soft Handover which is implemented in upper layer. FCGS will combat fast fading better than Soft Handover which has longer delays.

Fig. 3.9 Fast cell group
selection scheme topology

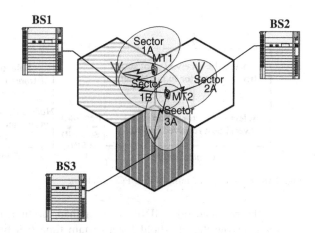

When deployed in sectoring system, FCGS will choose the best sector group to serve the MT. We will illustrate intra-NodeB FCGS, inter-NodeB FCGS in sectoring cellular system (ref. Fig. 3.9) and FCGS in non-sectoring respectively. Detailed description will be given in the following subsections.

3.2.1.1 Intra-NodeB Fast Cell Group Selection

In a sectoring system, when the MT locates at sector edges of the serving NodeB, it will process intra-NodeB Fast Cell Group Selection. As shown in Fig. 3.9, MT1 locates at the sector edge of sector 1A and sector 1B which belong to NodeB1. It will either select a sector between sector 1A and sector 1B to transmit downlink signal or select both of them according to instantaneous link quality.

In case of intra-NodeB, the downlink performance of the sector edge users is significantly degraded due to the features of sector antenna. Therefore an effective solution will be to increase the throughput of sector edge users by utilizing both sector antennas of the same NodeB, to transmit data simultaneously. The MT receives the simultaneous data from two sectors and combines them to resist the signal degradation. The problem of the synchronization can be ignored because both sector antennas belong to the same NodeB, the synchronization can be achieved easily. Compared to FCS scheme, the intra-NodeB FCGS can fully use the transmit diversity to further improve the sector edge user performance by choosing two sectors to transmit signals, when the link condition is unfavorable. Compared to softer handover, the FCGS is "faster" to resist the fast fading. The flow chart of intra-NodeB FCGS is indicated in Fig. 3.10.

The detailed implementation steps of intra-NodeB FCGS are shown as following (in the scenario of 3GPP 3G/E3G systems):

(1) MT measures the pilot signal's SINR of Common Pilot Channel (such as CPICH in HSDPA) of each sector. When the MT is located in the cell edge,

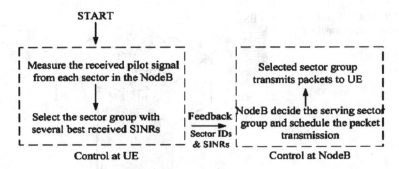

Fig. 3.10 Flow chart of intra-NodeB FCGS scheme

i.e. the received pilot SINR is lower than the threshold (set by system) and keeps below the threshold for a certain time (set by system), it comes into Intra-NodeB FCGS situation;

(2) MT selects a group of best (with largest pilot signal's SINRs) sectors as candidate sector group during every update duration (such as every frame), and inform the selection result (include pilot SINRs of each sector and the sector IDs of the selected candidate sector group) by uplink control channel (such as DPCCH in HSDPA for FCS);

(3) NodeB decides the serving sector group for each MT according to the feedback from MT and the network factors such as resource usage and system load etc.;

(4) NodeB decides the resource of the selected sector group and the transmission signal of STBC for each sector, then transmits the data to the MT;

(5) MT receives the signal from the sector group and decodes them.

3.2.1.2 Inter-NodeB Fast Cell Group Selection

The situation of inter-NodeB is little more complicated than intra-NodeB due to the problem of synchronization between NodeBs. Suppose the synchronization problem can be effectively solved, we can fully use the advantages of multi-NodeB transmit in this situation. The sketch map of inter-NodeB FCGS is shown as MT2 in Fig. 3.9. When MT2 locates at the common boundary of sector 1B of NodeB1, sector 2A of NodeB2 and sector 3A of NodeB3. It will select several sectors from these three sectors to transmit downlink signal according to instantaneous link condition. We define this process as inter-NodeB Fast Cell Group Selection.

The flow chart of inter-NodeB FCGS is similar to that illustrated in Fig. 3.10. The difference between inter and intra-NodeB FCGS is that inter-NodeB FCGS can select any sector of any NodeB as long as it has better link quality rather than only selecting sectors belonging to the same NodeB as intra-NodeB FCGS scheme. Because the sector edge region of one NodeB may be the main beam of the other

NodeB, inter-NodeB FCGS may achieve better performance than intra-NodeB FCGS with the cost of complexity for synchronization between NodeBs.

3.2.1.3 Non-Sectoring Fast Cell Group Selection

In non-sectoring system, the MT in the cell edge will select the best group of NodeBs to transmit. The communication mechanism between NodeBs can be the same as inter-NodeB FCGS of sectoring system. The flow chart is similar to that illustrated in Fig. 3.10.

It is worth mentioning that the sectoring inter-NodeB FCGS and non-sectoring system FCGS with STBC transmission scheme has the requirement of strict synchronization between NodeBs. RNC or SGW or other upper controller will be involved to schedule the sectors of different NodeBs to transmit STBC signals. If the strict synchronization cannot be ensured between BSs, Inter-NodeB FCGS with STBC transmission scheme will not be suitable, but intra-NodeB FCGS with STBC transmission scheme is still feasible.

3.2.2 Performance Analysis and Evaluation

In this section, we will compare the cell edge user performance of FCGS with Hard Handover, Soft Handover and FCS based on system level simulation. The user channel capacity (calculated by Shannon formula) will be adopted and the SINR will be analyzed (taking non-sectoring system as an example), which is the only difference among these schemes. And then we give the system-level simulation environment and results of sectoring intra-NodeB FCGS and non-sectoring system FCGS.

The user channel capacity is $C = B \log_2(1 + \frac{P_r}{Noise + I_{nonHO\_user} + I_{HO\_user}})$, where P_r is the receive power of the target user, I_{nonHO_user} is the interference from non-handover users of adjacent cells which is maintaining the same for four mentioned schemes and I_{HO_user} is the interference from handover users of adjacent cells. So, P_r and I_{HO_user} are the main factors that affect the user channel capacity.

We will compare P_r and I_{HO_user} of the four schemes respectively. For FCS scheme, MT may choose cells rather than Hard Handover scheme at the cell edge according to instantaneous link quality, so $P_{rFCS} \geq P_{rHHO}$; for Soft Handover scheme, MT choose best two cells in our simulation, and it has 100 ms delay which makes it incapable to combat fast fading. But we believe $P_{rSHO} \geq P_{rFCS}$ can be achieved considering Macro Diversity gain. For FCGS scheme, MT selects best two cells according to instantaneous link quality, so $P_{rFCGS} \geq P_{rSHO}$.

The interference of handover users in other cells of FCS ($I_{HO\_user\_FCS}$) is statistically the same as that of Hard Handover ($I_{HO\_user\_HHO}$), when the interfering user does not choose the target cell. But when the interfering user chooses

Table 3.2 Simulation parameters for FCGS

Simulation parameters	Settings
Bandwidth	10 MHz
Number of subcarriers	601
Subcarriers spacing	15 kHz
Cell layout	Hexagonal grid, 19 cell sites, 3 sectors per site
Cell radius	500 m
Minimum distance between MT and cell site	35 m
Antenna pattern	70-degree sectored beam
Distance dependent path loss	$128.1 + 37.6\log_{10}(r)$
Penetration loss	20 dB
Node B antenna gain plus cable loss	14 dBi
Noise figure	5 dB
Max Node B transmit power	43 dBm
Shadowing standard deviation	8 dB
Shadowing correlation between sectors/cells	1/0.5
Multipath delay profile	Typical urban
MT speed	3 km/h ($f_D = 5.55$ Hz)
Number of received antennas	1
Hard handover period	100 ms
Soft handover period	100 ms
Control delay in fast cell selection/fast cell group selection	5 ms
Frequency reuse factor	1
Subcarriers per user	20

the target cell, the interference will increase sharply. In this situation, some resource management methods are needed to mitigate co-channel interference from the same cell. For Soft Handover ($I_{HO\_user\_SHO}$) and FCGS scheme ($I_{HO\_user\_FCGS}$), choosing two cells to simultaneously transmit signals to MT will induce additional interferences, we use half transmit power to reduce interference and apply advance transmit technology to get diversity gain.

Based on OFDM technologies, suppose that the system is in full traffic, the frequency reuse factor is 1 and perfect synchronization can be guaranteed between NodeBs. Table 3.2 gives the parameters used in the simulations and they basically follow the simulation conditions described in [37, 38].

In the system-level simulation analysis, the intercell interference is considered. The simulation result is given in Figs. 3.11 and 3.12 and the target user in the simulation only focuses on cell edge users which can indicate the performance increment more distinctly.

Figure 3.11 shows the CDF of sector edge user channel capacity of sectoring system using FCS, Soft Handover scheme and Intra-NodeB FCGS scheme. The FCGS scheme in this simulation uses STBC to transmit downlink signals to the target user. So, the FCGS can get additional STBC transmit diversity. We can see from Fig. 3.11 that the sector edge user's performance of FCS and Soft

Fig. 3.11 Intra-NodeB fast cell group selection scheme

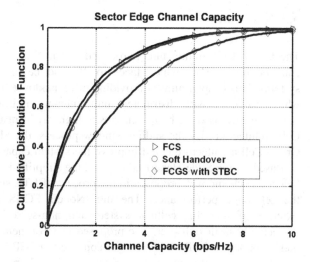

Fig. 3.12 Non-sectoring fast cell group selection scheme

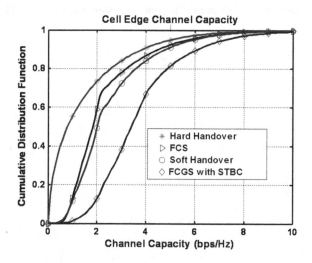

Handover degrade rapidly, but Intra-NodeB FCGS scheme can achieve higher user channel capacity and can have up to 60 % of gain compared to FCS and Soft Handover.

Figure 3.12 indicates the CDF of the cell edge user channel capacity using Hard Handover, FCS, Soft Handover and FCGS with STBC in non-sectoring system. We can see that the cell edge user performance by using FCGS in non-sectoring system is also much better than that of FCS, Soft Handover and Hard Handover scheme. The performance increment is obvious.

3.3 Summary

The Group Cell architecture introduced in the first chapter is a novel cellular construction method, which faces to 4G and beyond mobile communication systems. This chapter aims at providing fitful handover scheme for the Group Cell which can cope with distributed multi-antenna environments. Slide Handover scheme is proposed for both scenarios in highway infrastructure and urban Group Cell architecture. By the sliding window process, the Slide Handover scheme in Group Cell architecture could improve the system capacity by 10–30 %.

Based on the Slide Handover scheme, the simplified version—Fast Cell Group Selection scheme was proposed in the process of LTE standardization to improve the cell edge performance. The intra-NodeB FCGS and inter-NodeB FCGS schemes of sectoring cellular system and non-sectoring system are analyzed respectively in this chapter. The proposed FCGS scheme attracted many companies' supports in 3GPP LTE and got approved in 3GPP TR 25.814. Furthermore, the FCGS is actually the predecessor of CoMP with more than one antenna in the same base station or different base stations, so as the Group Cell architecture.

Chapter 4
Power Allocation of Group Cell System

Power Control (PC) refers to the adjustments of transmission power level based on certain criteria and it plays an important role in exploiting the capacity potential [39]. The power control transmitted by the radio base stations has been proposed to reduce the output power level and thus to reduce interference and increase the life time of the equipment. The reduction of interference implies some improvements of QoS or allows increasing the capacity of a wireless cellular system [40].

The downlink Group Cell scenarios need the cooperative transmission in multiple cells, which means one MT may be served by many different antenna units. The downlink Group Cell transmission implies dynamic coordination among multiple cells. That is, dynamic coordination among coordinated transmission points (CTPs) is conducted while not exceeding the power constraint of each CTP. It is necessary to jointly optimize the subcarrier resource allocation and the effective power control among the cooperative set to fully exploiting the capacity potential [41]. If they are not solved effectively, the performance benefits of Group Cell system will not be gained as expected. In this chapter, we consider the downlink power allocation problem based on the criterion of inter-cell interference coordination and maximum system capacity.

4.1 Downlink Resource Scheduling for ICIC

In OFDM systems, the subcarriers with the same frequency are reused among the different cells, it may bring into the additional inter-cell interference [42]. To be more specific, interference in OFDM-based systems arises when the same frequency resources are used in neighbor cells. For example, when two users are in different cells, and use the same frequency block simultaneously, then the signal to interference rate (SIR) associated with these blocks may drop to low level, resulting in a lower resource utilization and poor performance. Although for

X. Tao et al., *Group Cell Architecture for Cooperative Communications*,
SpringerBriefs in Computer Science, DOI: 10.1007/978-1-4614-4319-3_4,
© The Author(s) 2012

OFDM systems, subcarriers are orthogonal for users in intra-cell, it can effectively avoid intra-cell interference. However, the subcarriers with same frequency, reused among different cells may bring additional inter-cell interference. Moreover, for Group Cell scenarios, it needs cooperated transmission in the downlink, which means a user may be served by many different antenna units. So optimizing the subcarriers resources in downlink is significantly important for Group Cell technology [43].

On the basis of cooperation among antenna units and eNodeB, both the effective subcarrier resource allocation and adaptive power allocation methods needs to be proposed to optimize the subcarriers resources [44]. Such cooperation relationship selects the optimal subcarriers, establishes balanced SIR for co-frequency subcarriers, and also reduces inter-cell interference. The proposed subcarrier allocation in Group Cell is based on the gain interference ratio (GIR). By means of cooperation among antenna units and eNodeBs, the dynamic power allocation is also taken into consideration, which tries to establish a balanced signal to interference ratio (SIR) among cells. By this way, it enables to mitigate inter-cell interference, and improves system performance.

4.1.1 GIR-Based Subcarrier Resource Allocation

Fixed subcarrier allocation method is a traditional subcarrier allocation method in OFDM system. This method includes two types of subcarrier allocations namely, group and extended. These types allocate either group or interval subcarriers to each user, respectively [45].

The fixed method neglects the state of subcarriers, and may cause additional co-frequency subcarriers interference, which degrades the user throughput. By this analysis, a novel subcarrier allocation method is presented for Group Cell, which is based on GIR.

In OFDM system, the downlink GIR of a subcarrier can be defined as [46]:

$$C = \frac{S}{I} = \frac{\sum_{k} G_k}{N + \sum_{m} G_m P_m} \tag{4.1}$$

where $\sum_{k} G_k$ is the sum of kth subcarriers' channel gain (taking path loss, shadow fading and fast fading into consideration), N is the noise power, G_m is the channel gain from antenna unit m to the user, where antenna unit m belongs to other cells. P_m is the transmit power of antenna unit m, and $\sum_{k} G_k P_k$ is the sum of interference power from other cells.

As mentioned earlier that the proposed subcarrier allocation method is based on the gain interference ratio (GIR) from AUs to users. Moreover, all the available subcarriers are allocated according to GIR, and the process is given as following:

Set up a GIR matrix for subcarrier allocation, as shown in matrix (4.2), the row vectors denote the user flag, while the column vectors denote the carrier flags. In matrix (4.2), c_{ij} denotes the GIR value of user i on the subcarrier j, and it is assumed that each user uses the kth group of consecutive subcarriers.

$$\begin{bmatrix} c_{11} & \cdots & c_{1(j+1)} & \cdots & c_{1(j+k)} & \cdots & c_{1n} \\ \vdots & & \vdots & & \vdots & & \vdots \\ c_{i1} & \cdots & c_{i(j+1)} & \cdots & c_{i(j+k)} & \cdots & c_{in} \\ \vdots & & \vdots & & \vdots & & \vdots \\ c_{m1} & \cdots & c_{m(j+1)} & \cdots & c_{m(j+k)} & \cdots & c_{mn} \end{bmatrix} \quad (4.2)$$

Firstly, choose the largest elements in GIR matrix. If the largest elements are in the ith row and the $(j+1)$th $\sim (j+k)$th column, allocate these subcarriers to the user i. Then the elements in the ith row and the $(j+1)$th $\sim (j+k)$th column should be cleared, which means that the user i and the subcarriers $(j+1)$th $\sim (j+k)$th may not be participating in the next allocation process. If the subcarriers are released by users, then they can be considered in the next allocation process.

4.1.2 Balanced SIR-Based Power Allocation

For OFDM systems, the water-filling algorithm is usually considered for subcarrier allocation. Such allocation method is based on the channel state information (CSI). The CSI is suitable for a user then power should be increased otherwise it should be decreased. When this scheme is used in a single cell, it can effectively avoid intra-cell interference because the subcarriers are orthogonal for users. So the capacity of the system can be greatly improved. But for multicell scenarios, it may cause serious inter-cell co-frequency interference for other users because the subcarriers with the same frequency are reused among different cells.

According to inter-cell interference coordination in 3GPP LTE [47], power allocation schemes that allocate partial power to cell-center and full power to cell-edge users, are called as the fixed power allocation in this book. For this power allocation method, it can reduce the interference caused by users in cell-center, but on the other hand, it degrades the performance for these users. In addition, full power in cell-edge, may cause additional interference to users in the neighboring cells. So the power allocation method should not only fulfill the service requirement for intra-cell users, but also needs to mitigate inter-cell interference to users in other cells.

In order to optimize power allocation in inter-cell coordination schemes, a new power control algorithm is considered to adjust the power allocation in subcarriers, which is based on the cooperation among different antenna units and access points in Group Cell.

By means of cooperation relationship in Group Cell, the inter-cell interference information can be exchanged among antenna units and eNodeBs. So the power allocation can be dynamically adjusted among different antenna units and eNodeBs, the proposed power control algorithm aims to establish a balanced SIR for co-frequency subcarriers among inter-cells, which not only can adjust the power optimality, but also can mitigate inter-cell interference and improve system performance. This algorithm is introduced as follows.

According to the power ratio (PR), users in intra-cell are divided into two parts: cell-center users and cell-edge users. Specifically, the PR is defined as the ratio of useful power and co-frequency interference from co-frequency subcarriers in other cells. If the PR is higher than a standard value, such user is classified as cell-center user, otherwise the user is classified as cell-edge user. In order to mitigate the interference, subcarriers occupied by cell-center users are allocated with lower power. On the other hand, subcarriers occupied by cell-edge users are allocated with higher power, which can improve the performance in poor channel conditions. The target SIR must be set for different users, which is based on the service types and CSI. When all the parameters get initiated, the power adjustment will be started using power control algorithm, making it possible for the users' SIR to match the target values, and keeps SIR balanced among antenna units and eNodeBs. The proposed power control algorithm is in a form of iterations, and the main iterative process is that:

The iterative power is computed by the power iterative equation, and the SIR under such power can be derived by the SIR equation. Then, the next iterative power is updated by power iterative equation, and the new SIR under the iterative power is updated by SIR equation. If such SIR approximates to the target value, iteration should be stopped and power output should be generated, otherwise the iterative process should be continued until current SIR approximates to the target value.

For the maximum power constrain in power allocation, if the iterative power goes beyond the maximum power, and the current SIR still does not reach the target value, Such user should be removed and its initial parameters and priorities should be reset [48].The new iterative process should begin again considering single co-frequency subcarrier as an example, the steps of this iterative algorithm in Group Cell are given as follows:

(1) Set the initial parameters, such as the initial subcarriers power $p_{ij}^{(0)}$ from antenna unit j to user i, the target SIR γ_i^T, and the noise power v_i.

(2) According to the following power iterative equation, compute the modified power $p_{ii}^{(n+1)}$ in the next iteration $(n \geq 0)$:

$$p_{ii}^{(n+1)} = p_{ii}^{(n)} \cdot \frac{\gamma_i^T}{\gamma_i^{(n)}} \qquad (4.3)$$

(3) Compute $\gamma_i^{(n+1)}$ when the power is $p_{ii}^{(n+1)}$:

$$\gamma_i^{(n+1)} = \frac{g_{ii} p_{ii}^{(n+1)}}{\sum\limits_{j=1,j\neq i}^{N} g_{ij} p_{ij}^{(n+1)} + v_i} \qquad (4.4)$$

(4) If $\left| \gamma_i^{(n+1)} - \gamma_i^T \right| \leq \varepsilon$, output $p_{ii}^{(n+1)}$ and stop. Else, go to the next step.

(5) If $p_{ii}^{(n+1)} \leq p_{\max}$, $n = n + 1$, then go back to step (2). Else, remove the user with the minimum SIR, reset the priorities and go back step (1).

4.1.3 Subcarrier Optimization Algorithm

On the basis of the above analysis, the subcarrier resources can be optimized from two aspects, which respectively are subcarrier allocation and power allocation. According to subcarrier allocation based on GIR and power allocation based on balanced SIR, the subcarriers resources can be optimized for Group Cell, which is written as Scheme 1, and its process is given as follows:

Step 1: Confirm the serving access point according to the channel gain from antenna units to users.

Step 2: Classify all the intra-cell users into cell-edge users and cell-center users by the PR.

Step 3: Acquire the sum of channel gains on current subcarriers and the co-frequency interferences from other cells to current users. Then compute the GIRs for different users and available orthogonal subcarriers.

Step 4: Allocate the subcarriers according to the priority in line vectors. The larger the GIR value is, the higher the priority is.

Step 5: By means of the proposed optimizing power allocation algorithm in Group Cell, adjust the power in allocated subcarriers, and dynamically adjust SIR to be balanced among co-frequency subcarriers.

4.1.4 Performance Analysis and Evaluation

In order to compare the performance for the proposed subcarrier optimization method, simulation is performed in the Group Cell scenario, and the basic

Table 4.1 Simulation parameters for group cell scenario

Parameters	Settings
Channel environment	Macro-cell Hata model
Carrier frequency	2 GHz
Bandwidth	10 MHz
Distance of subcarrier	15 kHz
Sampling frequency	1.92 MHz
CP length	7.29/14 μs/sample
FFT size	1,024
The number of carriers	600
The number of cells	7
Cell radius	1 km
Channel model	Typical urban (TU)
Maximum power in BS	43 dBm

simulation parameters are referenced from LTE, and some of them are shown in Table 4.1 [49, 50].

Four schemes for subcarriers resource optimization are compared in simulation, which respectively are as follows:

(1) Scheme 1

Group Cell: Subcarrier allocation based on GIR
Group Cell: Power allocation based on balanced SIR

(2) Scheme 2

Fixed subcarrier allocation
Group Cell: Power allocation based on balanced SIR

(3) Scheme 3

Group Cell: Subcarrier allocation based on GIR
Fixed power allocation

(4) Scheme 4

Fixed subcarrier allocation
Fixed power allocation

In Scheme 1, the subcarriers optimization methods are based on cooperation among different antenna units and eNodeBs, which are respectively subcarrier allocation based on GIR, and power allocation based on balanced SIR. On the basis of Group Cell, the state of co-frequency subcarriers are cooperated among cells, which enables to control the inter-cell interference, subcarriers are then selected and power is allocated. In Scheme 2, it takes fixed subcarrier allocation (allocates a group of adjacent subcarriers or interval subcarriers to users), while using dynamic power allocation based on balanced SIR. In Scheme 3, it takes subcarrier allocation based on GIR, while taking fixed power allocation (partial power in cell-center,

Fig. 4.1 Throughput comparison

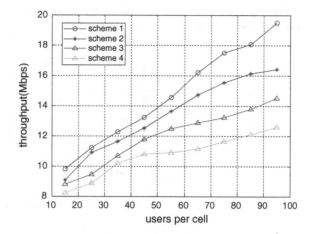

and full power in cell-edge). In Scheme 4, as a traditional scheme, both fixed subcarrier allocation and fixed power allocation are taken.

Figure 4.1 shows the throughput comparison of different subcarriers resource optimization schemes. It can be seen that as the number of users increases so do the cells throughout. Especially, when users per cell are the same with the above mentioned four schemes, the throughput in Scheme 1 is increased as compared to Scheme 2, Scheme 3, and Scheme 4 which illustrates that the subcarriers resources are optimized by means of Group Cell and the performance has been improved.

Scheme 2 is better than Scheme 3, as each scheme uses different optimization approaches. One approach is in power allocation, whose process is dynamic, and the power on each subcarrier is adjusted to keep a balanced SIR among cells. The other approach is in subcarrier allocation, which is a static process, and which searches for the largest GIR as the objective subcarriers. From such comparisons, it can be seen that the power optimization plays an important role in subcarriers optimization process.

Scheme 4 uses a traditional approach for subcarriers, which lacks cooperation. Compared with Scheme 3 and Scheme 4, it shows that the improved throughput by utilizing the proposed subcarrier allocation method, whose performance is better than those in traditional ways.

Figure 4.2 compares the blocking probability with the change of number of users per cell, as per different schemes. From the graph, it can be seen that the blocking probability grows with the increase of users per cell.

Compared with Scheme 2 and Scheme 4, it reveals the performance brought by the proposed power allocation method. While the comparison with Scheme 3 and Scheme 4 it shows the performance gain brought by the proposed subcarrier allocation method. By optimizing subcarriers, the blocking probabilities are decreased compared to the traditional approaches. Moreover, the blocking probability decreases more by using the proposed power allocation method as compared to the proposed subcarrier allocation method.

Fig. 4.2 Blocking
probability comparison

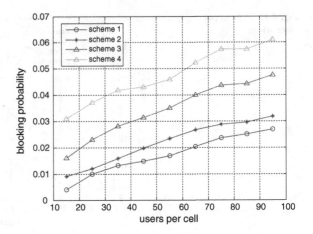

Fig. 4.3 Average data rates
comparison for cell-edge
users

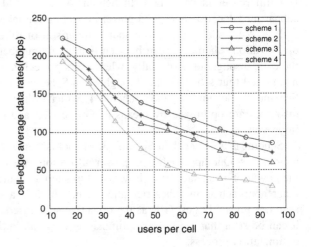

Specially, it jointly takes two subcarriers optimization ways in Scheme 1, and its blocking probability is reduced into the minimum among four schemes.

Figure 4.3 shows the comparison for average data rates in cell-edge. Like throughput of cell edge users, it can be seen that the average data rates are increased mostly by Scheme 1. In addition, Scheme 2 and Scheme 3 are also with good performances. As a traditional approach, the improvement of data rates brought by Scheme 4 is not very significant.

From this comparison, it can be found that the data rates at cell-edge users are greatly increased by the proposed schemes, which proves that such schemes with Group Cell can improve the performance at cell-edge.

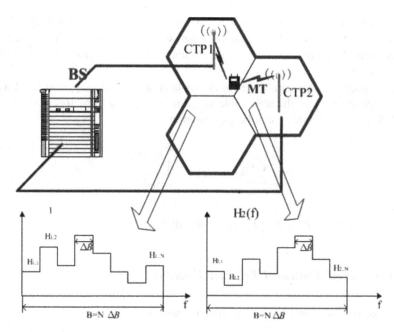

Fig. 4.4 Two CTPs coordinately allocate their transmission power

4.2 Downlink Power Allocation for Maximizing System Capacity

In this section, we consider the Shannon capacity of frequency-selective fading channel with multiple CTP, as depicted in the Fig. 4.4, in which two CTPs are considered which transmit signals to a certain MT. Each CTP has all the information that the MT require but the power constraint at each CTP is assumed to be different as $P_i, i = 1, 2$. For the sake of convenience, we assume the channel response $H_i(f)$ is slowly time-varying and block-fading, so that the frequency can be approximately divided into several flat narrow-band subchannels. The CTPs then coordinately allocate their power over each subchannel so as to achieve the channel capacity. This approach naturally leads to multitone techniques such as OFDM [51].

The channel state information (CSI) is assumed to be available at both the two CTP and the receiver, and can be exchanged reliably and promptly between the two CTPs. Assuming the total bandwidth is B, which is divided into N subchannels of bandwidth ΔB. The MT can optimally receive its information at each subchannel with maximal-ratio combining (MRC). Thus, the received signal-to-noise ratio (SNR) on the jth subchannel is given by

$$SNR_j = \frac{\sum_{i=1}^{2} P_{i,j}|H_{i,j}|^2}{N_0 \Delta B} \tag{4.5}$$

where $N_0 \Delta B$ represents the power of additive white Gaussian noise at the jth subchannel. $P_{i,j}$ denotes the allocated transmission power from CTP i to the jth subchannel; $H_{i,j}$ denotes the corresponding subchannel gain between CTP i and the jth subchannel.

The capacity of this parallel set of channels is the sum of rates associated with each subchannel. Thus, the achievable sum throughput (rate) which is given by the AWGN Shannon Capacity can be expressed as

$$R = \sum_{j=1}^{N} \Delta B \log_2 \left(1 + \frac{\sum_{i=1}^{2} P_{i,j} |H_{i,j}|^2}{N_0 \Delta B}\right) \tag{4.6}$$

Subject to $\sum_{j=1}^{N} P_{i,j} = P_i$, $i = 1, 2$, $P_{i,j} \geq 0$ for all i, j.

4.2.1 Optimal Transmit Power Allocation

The objective function (4.6) can be transformed as

$$R' = R/\Delta B = \sum_{j=1}^{N} \log_2 \left(1 + \frac{\sum_{i=1}^{2} P_{i,j} |H_{i,j}|^2}{N_0 \Delta B}\right) \tag{4.7}$$

We search for the optimal cooperative joint power allocation by approaching the following optimization problem

$$P_i^* = (P_{i,1}^*, \ldots, P_{i,N}^*) = \arg \max_{P_i \in \Omega^N} R', \quad i = 1, 2 \tag{4.8}$$

where $\Omega^N = \{P_i | \forall j \in (1, \cdots N), 0 \leq P_{i,j} \leq P_i, \sum_{j=1}^{N} P_{i,j} = P_i\}$ is the feasible set, since Ω^N is a closed and bounded set and $R' : \Omega^N \rightarrow \mathbb{R}$ is continuous, thus, Eq. (4.7) have a solution [52].

We let $\gamma_{i,j} = (P_i |H_{i,j}|^2)/(N_0 \Delta B)$ and $x_{i,j} = P_{i,j}/P_i$. It is indicated that $\gamma_{i,j} = (P_i |H_{i,j}|^2)/(N_0 \Delta B)$ is the SNR associated with CTP i over the jth subchannel when assuming it is allocated the entire power P_i, and $x_{i,j}$ represents the power allocation ratio. Thus the objective function (4.7) can be transformed as

$$R' = R/\Delta B = \sum_{j=1}^{N} \log_2 \left(1 + \sum_{i=1}^{2} \gamma_{i,j} x_{i,j}\right) \tag{4.9}$$

Lemma 1 In order to maximize the achievable throughput, only one of the N subchannels ought to be transmitted together by the two CTP, the others should be divided into two parts and transmitted by respective single CTP.

Since the logarithm is monotonically increasing, by $\sum_{j=1}^{N} x_{i,j} = 1$, $i \in \{1, 2\}$, the objective function (4.9) which incorporates the power constraint can be defined as

$$f(x_{1,1}, x_{2,1}, \cdots x_{1,j}, x_{2,j}, \cdots x_{1,N-1}, x_{2,N-1})$$
$$+ \gamma_{2,N}(1 - x_{2,1} \cdots - x_{2,j} \cdots - x_{2,N-1})] \tag{4.10}$$

Next, we will specifically analyze the features of the above function which contains 2N−2 variables.

Firstly, we prove function (4.10) has no extreme point. We denote $A_j = (1 + \gamma_{1,j} x_{1,j} + \gamma_{2,j} x_{2,j})$, $j \in (1 \cdots N - 1)$, and obtain the Hessian matrix for arbitrary point in domain $\nabla^2 f$ as follows

$$\nabla^2 f = \begin{bmatrix} \frac{\partial^2 f}{\partial x_{1,1}^2} & \frac{\partial^2 f}{\partial x_{1,1} \partial x_{2,1}} & \cdots & \frac{\partial^2 f}{\partial x_{1,1} \partial x_{1,n-1}} & \frac{\partial^2 f}{\partial x_{1,1} \partial x_{2,n-1}} \\ \frac{\partial^2 f}{\partial x_{2,1} \partial x_{1,1}} & \frac{\partial^2 f}{\partial x_{2,1}^2} & \cdots & & \\ \vdots & & \ddots & & \vdots \\ \frac{\partial^2 f}{\partial x_{1,n-1} \partial x_{1,1}} & & \cdots & \frac{\partial^2 f}{\partial x_{1,n-1}^2} & \\ \frac{\partial^2 f}{\partial x_{2,n-1} \partial x_{1,1}} & & \cdots & & \frac{\partial^2 f}{\partial x_{2,n-1}^2} \end{bmatrix} \tag{4.11}$$

Then, we get the following regular equations

$$\frac{\partial^2 f}{\partial x_{1,1}^2} = -2\gamma_{1,1}\gamma_{1,N}A_2A_3 \cdots A_{N-1},$$

$$\frac{\partial^2 f}{\partial x_{2,1}^2} = -2\gamma_{2,1}\gamma_{2,N}A_2A_3 \cdots A_{N-1}, \tag{4.12}$$

$$\frac{\partial^2 f}{\partial x_{1,1} \partial x_{2,1}} = \frac{\partial^2 f}{\partial x_{2,1} \partial x_{1,1}} = -(\gamma_{1,1}\gamma_{2,N} + \gamma_{2,1}\gamma_{1,N})A_2A_3 \cdots A_{N-1}.$$

From the above equations, it is seen that the first-order determinant is $-2\gamma_{1,1}\gamma_{1,N}A_2A_3 \cdots A_{N-1} < 0$, and the second-order principal minor determinant is obtained as follows

$$\frac{\partial^2 f}{\partial x_{1,1}^2} \cdot \frac{\partial^2 f}{\partial x_{2,1}^2} - \frac{\partial^2 f}{\partial x_{1,1} \partial x_{2,1}} \cdot \frac{\partial^2 f}{\partial x_{2,1} \partial x_{1,1}} = -(\gamma_{1,1}\gamma_{2,N} - \gamma_{2,1}\gamma_{1,N})^2$$
$$\cdot (A_2A_3 \cdots A_{N-1})^2 < 0 \tag{4.13}$$

It is emphasized that Eq. (4.13) is negative due to the possibility of $|H_{1,1}|^2|H_{2,N}|^2 = |H_{2,1}|^2|H_{1,N}|^2$ is almost zero in practical system model. Thus the Hessian $\nabla^2 f$ is indefinite, which means function (4.10) has no extreme points, so the maximum value must be achieved on the domain boundary [29]. In other words, one of the 2N−2 variables is zero due to the closed and bounded set of $x_{i,j}$, and function (4.10) should be degraded to $2N-1$ dimensions. That is $A_1 = (1 + \gamma_{1,1}x_{1,1} + \gamma_{2,1}x_{2,1})$ has to be degraded to $A_1^\Delta = (1 + \gamma_{i,1}x_{i,1})$, $i = 1\ or\ 2$.

Similarly, for arbitrary $A_j = (1 + \gamma_{1j}x_{1j} + \gamma_{2j}x_{2j})$, we still have the following equation.

$$\frac{\partial^2 f}{\partial x_{1,j}^2}\frac{\partial^2 f}{\partial x_{2,j}^2} - \frac{\partial^2 f}{\partial x_{1,j}\partial x_{2,j}}\frac{\partial^2 f}{\partial x_{2,j}\partial x_{1,j}} = -(\gamma_{1,j}\gamma_{2,N} - \gamma_{2,j}\gamma_{1,N})^2$$

$$\cdot (A_1 \cdots A_{j-1}A_{j+1} \cdots A_{N-1})^2 < 0 \qquad (4.14)$$

Hence, in order to obtain the maximum value, all $A_j = (1 + \gamma_{1,j}x_{1j} + \gamma_{2,j}x_{2,j})$, $j \in \{1, \cdots, N-1\}$ have to be degraded to $A_j^\Delta = (1 + \gamma_{i,j}x_{i,j})$, $i = 1\ or\ 2$, $j \in \{1, 2, \cdots, N-1\}$.

In the end, a certain $A_n(n \in \{1 \cdots N\})$ is left, which contains two additive variables $(1 + \gamma_{1,n}x_{1,n} + \gamma_{2,n}x_{2,n})$ due to the constraint $\sum_{j=1}^{N} x_{i,j} = 1$, $i \in \{1,2\}$, the other A_j are degraded to the form $A_j^\Delta = (1 + \gamma_{i,j}x_{i,j})$, $i = 1\ or\ 2$, thus function (4.10) has been turned into strictly convex.

Remark 1 Based on the above analysis, the optimal power allocation mode is that only one of the N-subchannels (a certain A_n) should be transmitted together by the two CTP; the others, according to the different subchannel gains, should be separated into two parts ($x_{i,j}$ or $x_{2,j}$) and transmitted by respective single CTP. However, which variable ($x_{1,j}$ or $x_{2,j}$) should be removed depends on the specific subchannel gains. In the next part we solve the degraded convex problem, and obtain the general closed form solution. Based on the solution a novel joint power allocation named as joint waterfilling (Jo-WF) is presented.

Lemma 2 The optimal power allocation for each subchannel turns out to take the form of traditional water-filling, and also combined with a cooperative feature.

Proof First, we divide the N subchannels into two parts. Without loss of generality, suppose that the nth subchannel is the common subchannel which is transmitted together by the two CTP; some K subchannels are transmitted by CTP 1; the other M subchannels are transmitted by CTP 2. Then the objective function (4.9) can be expressed as

$$R' = \sum_{k=1}^{K} \log_2(1 + \gamma_{1,k}x_{1,k}) + \log_2(1 + \gamma_{1,n}x_{1,n} + \gamma_{2,n}x_{2,n})$$

$$+ \sum_{m=1}^{M} \log_2(1 + \gamma_{2,m}x_{2,m})$$

Subject to $N = K + M + 1$

$$(4.15)$$

By Lagrange dual function we have

$$\Gamma(\lambda_1, \lambda_2) = \sum_{k=1}^{K} \log_2(1 + \gamma_{1,k}x_{1,k}) + \log_2(1 + \gamma_{1,n}x_{1,n} + \gamma_{2,n}x_{2,n})$$

$$+ \sum_{m=1}^{M} \log_2(1 + \gamma_{2,m}x_{2,m}) + \lambda_1 \left(\sum_{k=1}^{K} x_{1,k} + x_{1,n} - 1\right) \qquad (4.16)$$

$$+ \lambda_2 \left(\sum_{m=1}^{M} x_{2,m} + x_{2,n} - 1\right)$$

We get the following solution by partial derivation on (4.16). First, we obtain the two cutoff value λ_1 and λ_2 as follows

$$\begin{cases} \dfrac{1}{\lambda_1} = \dfrac{1 + \gamma_{1,n}\left(1 + \sum_{k=1}^{K} \frac{1}{\gamma_{1,k}}\right) + \gamma_{2,n}\left(1 + \sum_{m=1}^{M} \frac{1}{\gamma_{2,m}}\right)}{\ln 2 \cdot N\gamma_{1,n}} \\[4mm] \dfrac{1}{\lambda_2} = \dfrac{1 + \gamma_{1,n}\left(1 + \sum_{k=1}^{K} \frac{1}{\gamma_{1,k}}\right) + \gamma_{2,n}\left(1 + \sum_{m=1}^{M} \frac{1}{\gamma_{2,m}}\right)}{\ln 2 \cdot N\gamma_{2,n}} \end{cases} \qquad (4.17)$$

It is indicated that the two cutoff value s are associated by the common subchannel n, and each cutoff value contains all the subchannel state information. The cutoff values can be obtained only by perfect CSI exchanging between the two CTP.

Then, according to the derived cutoff value s, the closed form solutions of Jo-WF power allocation are obtained as follows

$$\begin{cases} x_{1,k} = \frac{1}{\ln 2 \cdot \lambda_1} - \frac{1}{\gamma_{1,k}}, & x_{1,n} = 1 + \sum_{k=1}^{K} \frac{1}{\gamma_{1,k}} - \frac{K}{\ln 2 \cdot \lambda_1} \\[4mm] x_{2,m} = \frac{1}{\ln 2 \cdot \lambda_2} - \frac{1}{\gamma_{2,m}}, & x_{2,n} = 1 + \sum_{m=1}^{M} \frac{1}{\gamma_{2,m}} - \frac{M}{\ln 2 \cdot \lambda_2} \end{cases} \qquad (4.18)$$

It is indicated that the optimal power allocation which we name Jo-WF turns out to take the form of traditional water-filling and also have a cooperation property, which means the optimal power allocation ratio $x_{1,k}$ correlates with not only $\gamma_{1,k}, k \in (1, \cdots K)$, but also $\gamma_{2,m}, m \in (1, \cdots, M)$. The solution can be obtained only by perfect cooperation (CSI exchange) between the two CTP.

Note: Eq. (4.18) shows that the amount of power allocated for a given subchannel is $1/(\ln 2 \cdot \lambda_1) - 1/\gamma_{1,k}$ or $1/(\ln 2 \cdot \lambda_2) - 1/\gamma_{2,m}$ (except for the common subshannel), so the intuition behind the Jo-WF is to take advantage of good channel condition: when channel conditions are good, more power and a higher data rate should be sent over the channel. Considering that $\ln 2 \cdot \lambda_1$ and $\ln 2 \cdot \lambda_2$ are the two cutoff values, the solution in (4.19) should be satisfied by $P_{i,j} \geq 0$, in this case, we have $P_{i,j}^* = [P_{i,j}]^+ = \max\{0, P_{i,j}\}$, which means if the channel condition

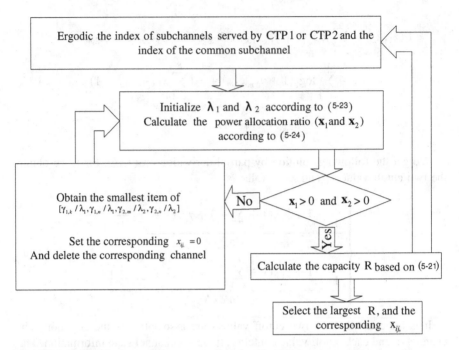

Fig. 4.5 Iterative joint water-filling power allocation algorithm

falls below the cutoff values the channel is not used. So an iterated method should be used to find the appropriate cutoff values.

Hence, we present a new iterative joint waterfilling power allocation algorithm as depicted in Fig. 4.5 so as to seek the two optimal cutoff values as well as evaluate the solutions numerically.

Now extend the two CTP case to arbitrary K-CTP case, naturally we obtain $\lambda_1, \cdots \lambda_k$ as follows

$$
\begin{cases}
\dfrac{1}{\lambda_1} = \dfrac{1 + \gamma_{1,n}\left(1 + \sum\limits_{k_1=1}^{K_1} \frac{1}{\gamma_{1,k_1}}\right) \cdots + \gamma_{k,n}\left(1 + \sum\limits_{k_k=1}^{K_k} \frac{1}{\gamma_{k,k_k}}\right)}{\ln 2 \cdot N\gamma_{1,n}} \\
\;\vdots \\
\dfrac{1}{\lambda_k} = \dfrac{1 + \gamma_{1,n}\left(1 + \sum\limits_{k_1=1}^{K_1} \frac{1}{\gamma_{1,k_1}}\right) \cdots + \gamma_{k,n}\left(1 + \sum\limits_{k_k=1}^{K_k} \frac{1}{\gamma_{k,k_k}}\right)}{\ln 2 \cdot N\gamma_{k,n}}
\end{cases} \tag{4.19}
$$

Numerical simulation process is presented in Fig. 4.5.

4.2.2 Sub-Optimal Power Allocation Scheme

To utilize this joint-waterfilling in practical system, an important issue is the complex transmitter and receiver design, such as variable-rate variable-power MQAM modulation and coding. In this part, we investigate a new constant-power joint-waterfilling scheme for a coordinated transmission system, in which two constant power levels are used across a properly chosen subset of subchannels. A rigorous worst-case performance bound of the constant-power joint-waterfilling is given based on the duality gap analysis. Furthermore, a low-complexity constant-power adaptation algorithm is also developed.

4.2.2.1 Duality Gap

A general form of a convex optimization problem is the following:

$$\min f_0(x)$$
$$s.t. f_i(x) \leq 0 \tag{4.20}$$

where $f_i(x), i = 1, 2, \ldots, m$ are convex functions. $f_0(x)$ is called the primal objective. The Lagrangian of the optimization problem is defined as

$$L(x, \lambda) = f_0(x) + \lambda_1 f_1(x) + \lambda_2 f_2(x) + \cdots + \lambda_m f_m(x) \tag{4.21}$$

where λ_i are nonnegative. The dual objective is defined as $g(\lambda) = \inf_x L(x, \lambda)$. It is easy to see that $g(\lambda)$ is a lower bound on the optimal $f_0(x)$

$$f_0(x) \geq f_0(x) + \sum_i \lambda_i f_i(x)$$
$$\geq \inf_z \left(f_0(z) + \sum_i \lambda_i f_i(z) \right) \geq g(\lambda) \tag{4.22}$$

Hence

$$g(\lambda) \leq \min_x f_0(x) \tag{4.23}$$

This is the lower bound which we will use to analyze the optimality of approximate joint-waterfilling schemes. The difference between the primal objective $f_0(x)$ and the dual objective $g(\lambda)$ is called the duality gap. In [53], the main result is when the primal problem is convex, the duality gap reduces to zero at the optimum, under some constraint qualifications. In other words, the optimal value of the primal objective may be obtained by maximizing the dual objective $g(\lambda)$ over nonnegative dual variable λ_i. Therefore, for convex problems, the lower bound is strictly tight.

4.2.2.2 Lower Bound of Joint-Waterfilling

First, maximizing the data rate is equivalent to minimizing its negative. The capacity is a concave function of power, so its negative is convex. The constraints of the objective are linear, and they are also convex. Associate dual variable λ_i, with the two CTP's power constraints, and $u_{i,j}$ with each of the positively con-straints on $x_{i,j}$, the Lagrangian $L(x_{i,j}, \lambda_i, u_{i,j})$ is then

$$
\begin{aligned}
L(x_{i,j}, \lambda_i, u_{i,j}) = & -\sum_{k=1}^{K} \log_2(1 + \gamma_{1,k}x_{1,k}) - \log_2(1 + \gamma_{1,n}x_{1,n} + \gamma_{2,n}x_{2,n}) \\
& -\sum_{m=1}^{M} \log_2(1 + \gamma_{2,m}x_{2,m}) + \lambda_1(\sum_{k=1}^{K} x_{1,k} + x_{1,n} - 1) \\
& + \lambda_2(\sum_{m=1}^{M} x_{2,m} + x_{2,n} - 1) + \sum_{k=1}^{K} u_{1,k}(-x_{1,k}) + u_{1,n}(-x_{1,n}) \\
& + \sum_{m=1}^{M} u_{2,m}(-x_{2,m}) + u_{2,n}(-x_{2,n})
\end{aligned}
\tag{4.24}
$$

The dual objective function $g(\lambda_i, u_{i,j})$ is the infimum of the Lagrangian over primal variable $x_{i,j}$. At the infimum, the partial derivation of the Lagrangian with respect to $x_{i,j}$ must be zero $\partial L / \partial x_{i,j} = 0$, without loss of generality, from which the joint-waterfilling condition is obtained as follows:

For CTP 1 we have

$$
\begin{aligned}
x_{1,k} + \frac{1}{\gamma_{1,k}} &= \frac{1}{\ln 2} \cdot \frac{1}{\lambda_1 - u_{1,k}}, k \in \{1, \ldots, K\}; \\
x_{1,n} + \frac{1}{\gamma_{1,n}} + \frac{\gamma_{2,n}}{\gamma_{1,n}}x_{2,n} &= \frac{1}{\ln 2} \cdot \frac{1}{\lambda_1 - u_{1,n}}.
\end{aligned}
\tag{4.25}
$$

For CTP 2 we have

$$
\begin{aligned}
x_{2,m} + \frac{1}{\gamma_{2,m}} &= \frac{1}{\ln 2} \cdot \frac{1}{\lambda_2 - u_{2,m}}, m \in \{1, \ldots, M\}; \\
x_{2,n} + \frac{1}{\gamma_{2,n}} + \frac{\gamma_{1,n}}{\gamma_{2,n}}x_{1,n} &= \frac{1}{\ln 2} \cdot \frac{1}{\lambda_2 - u_{2,n}}.
\end{aligned}
\tag{4.26}
$$

In convex optimization problem, the condition in (4.25) and (4.26), together with the constraints of the original primal problem, the positivity constraints on the dual variables, and the complementary slackness constraints, form the Karush–Kuhn–Tucker (KKT) condition, which is sufficient and necessary in this case. Specifically, according to the complementary slackness condition, the constraint of the original primal problem is satisfied with equality only when the dual variable associated with the inequality is greater than zero. In the joint-waterfilling prob-lem, this translates to the condition that $x_{i,j}$ is greater than zero if and only if $u_{i,j}$ is zero. In this case, $x_{i,j}^* = [x_{i,j}]^+ = \max\{0, x_{i,j}\}$, which means, if the channel condition falls below the cutoff threshold (waterlevel) the channel is not used. Substituting the joint waterfilling condition (4.25) and (4.26) into (4.24) gives the dual objective $g(\lambda_i, u_{i,k})$

$$
\begin{aligned}
g(\lambda_i, u_{i,k}) = & -\sum_{k=1}^{K} \log_2\left(\frac{1}{\ln 2} \cdot \frac{\gamma_{1,k}}{\lambda_1 - u_{1,k}}\right) - \log_2\left(\frac{1}{\ln 2} \cdot \frac{\gamma_{1,n}}{\lambda_1 - u_{1,n}}\right) \\
& -\sum_{m=1}^{M} \log_2\left(\frac{1}{\ln 2} \cdot \frac{\gamma_{2,m}}{\lambda_2 - u_{2,m}}\right) - \sum_{k=1}^{K} \frac{\lambda_1 - u_{1,k}}{\gamma_{1,k}} - \frac{\lambda_1 - u_{1,n}}{\gamma_{1,n}} \\
& -\sum_{m=1}^{M} \frac{\lambda_2 - u_{2,m}}{\gamma_{2,m}} - \lambda_1 - \lambda_2 + \frac{N}{\ln 2}
\end{aligned}
\tag{4.27}
$$

The dual objective is always convex, and it is a lower bound (Note that $g(\lambda_i, u_{i,j})$ is a lower bound, so $g(\lambda_i, u_{i,j})$ is an upper bound) to the primal objective for all nonnegative λ_i and $u_{i,j}$. In fact, the lower bound is tight when λ_i and $u_{i,j}$ achieve the optimum of the dual problem. Finding tightest λ_i and $u_{i,j}$ is equivalent to solving the original optimization problem, which is complicated. However, when dual variables associated with the primal variables are chosen via (4.25), (4.26) and (4.27), then a simple bound emerges. In this case, the duality gap, defined as the difference between the primal objective and the dual objective in joint-waterfilling, and denoted as

$$
\Gamma = \sum_{k=1}^{K} \frac{\lambda_1 - u_{1,k}}{\gamma_{1,k}} + \frac{\lambda_1 - u_{1,n}}{\gamma_{1,n}} + \sum_{m=1}^{M} \frac{\lambda_2 - u_{2,m}}{\gamma_{2,m}} + \lambda_1 + \lambda_2 - \frac{N}{\ln 2}
\tag{4.28}
$$

To express the gap exclusively in primal variable $x_{i,j}$, a suitable λ_i needs to be found. A small λ_i is desirable, because it makes the duality gap small. Recall that λ_i and $u_{i,j}$ need to be nonnegative, without loss of generality, the smallest nonnegative λ_i is then

$$
\lambda_1 = \max_k \left(\frac{1}{\ln 2} \cdot \frac{1}{x_{1,k} + \frac{1}{\gamma_{1,k}}}\right) = \frac{1}{\ln 2} \cdot \frac{1}{\min_k \left(x_{1,k} + \frac{1}{\gamma_{1,k}}\right)}
\tag{4.29}
$$

$$
\lambda_2 = \max_m \left(\frac{1}{\ln 2} \cdot \frac{1}{x_{2,m} + \frac{1}{\gamma_{2,m}}}\right) = \frac{1}{\ln 2} \cdot \frac{1}{\min_m \left(x_{1,m} + \frac{1}{\gamma_{1,m}}\right)}
\tag{4.30}
$$

Substituting the derived λ_i into the duality gap Γ gives:

$$
\Gamma = \frac{1}{\ln 2} \left[\sum_{k=1}^{K} \left(\frac{x_{1,k}}{\min\limits_{k}(x_{1,k} + 1/\gamma_{1,k})} - \frac{x_{1,k}}{(x_{1,k} + 1/\gamma_{1,k})} \right) + \frac{x_{1,n}}{\min\limits_{k}(x_{1,k} + 1/\gamma_{1,k})} \right.
$$

$$
- \frac{x_{1,n}}{x_{1,n} + (1 + x_{2,n}\gamma_{2,n})/\gamma_{1,n}} + \sum_{m=1}^{M} \left(\frac{x_{2,m}}{\min\limits_{m}(x_{2,m} + 1/\gamma_{2,m})} - \frac{x_{2,m}}{(x_{2,m} + 1/\gamma_{2,m})} \right) \quad (4.31)
$$

$$
\left. + \frac{x_{2,n}}{\min\limits_{m}(x_{2,m} + 1/\gamma_{2,m})} - \frac{x_{2,n}}{x_{2,n} + (1 + x_{1,n}\gamma_{1,n})/\gamma_{2,n}} \right]
$$

4.2.2.3 Constant-Power Joint-Waterfilling Algorithm

In [54], Chow discovered that as long as a correct frequency band is used, a constant power allocation has a negligible performance loss compared to true watefilling. In this section, based on the above duality analysis, we extend the existing constant-power waterfilling into the coordinated transmission system, and proposed a new constant-power joint-waterfilling scheme. Before we move on, a rigorous worst case performance of the constant-power joint-waterfilling is derived first using the derived gap bound. Consider the following class of constant-power allocation strategies, where under a joint-waterfilling cutoff threshold $1/(\ln 2 \cdot \lambda_i)$, all subchannels are allocated the same power ratio $x_{i,0}$, $i = 1, 2$

$$
x_{1,k} = \begin{cases} x_{1,0}, & \text{if } \dfrac{1}{\gamma_{1,k}} \le \dfrac{1}{\ln 2 \cdot \lambda_1} \\ 0, & \text{if } \dfrac{1}{\gamma_{1,k}} > \dfrac{1}{\ln 2 \cdot \lambda_1} \end{cases} ; \quad x_{2,m} = \begin{cases} x_{2,0}, & \text{if } \dfrac{1}{\gamma_{2,m}} \le \dfrac{1}{\ln 2 \cdot \lambda_2} \\ 0, & \text{if } \dfrac{1}{\gamma_{2,m}} > \dfrac{1}{\ln 2 \cdot \lambda_2} \end{cases} \quad (4.32)
$$

In this case, the duality gap in formula (4.31) becomes

$$
\Gamma = \frac{1}{\ln 2} \left[\sum_{k=1}^{K'} \left(\frac{x_{1,0}(\frac{1}{\gamma_{1,k}} - \min\limits_{k_0}\frac{1}{\gamma_{1,k_0}})}{(x_{1,0} + \min\limits_{k_0}\frac{1}{\gamma_{1,k_0}})(x_{1,0} + \frac{1}{\gamma_{1,k}})} \right) + \frac{x_{1,0}[\frac{1+x_{2,0}\gamma_{2,n}}{\gamma_{1,n}} - \min\limits_{k_0}\frac{1}{\gamma_{1,k_0}}]}{[x_{1,0} + \min\limits_{k_0}\frac{1}{\gamma_{1,k_0}}][x_{1,0} + \frac{1+x_{2,0}\gamma_{2,n}}{\gamma_{1,n}}]} \right.
$$

$$
\left. + \sum_{m=1}^{M'} \left(\frac{x_{2,0}(\frac{1}{\gamma_{2,m}} - \min\limits_{m_0}\frac{1}{\gamma_{2,m_0}})}{(x_{2,0} + \min\limits_{m_0}\frac{1}{\gamma_{2,m_0}})(x_{2,0} + \frac{1}{\gamma_{2,m}})} \right) + \frac{x_{2,0}[\frac{1+x_{1,0}\gamma_{1,n}}{\gamma_{2,n}} - \min\limits_{m_0}\frac{1}{\gamma_{2,m_0}}]}{[x_{2,0} + \min\limits_{m_0}\frac{1}{\gamma_{2,m_0}}][x_{2,0} + \frac{1+x_{1,0}\gamma_{1,n}}{\gamma_{2,n}}]} \right]
$$

$$
\le \frac{1}{\ln 2} \left[\sum_{k=1}^{K'} \left(\frac{\frac{1}{\gamma_{1,k}}}{x_{1,0} + \frac{1}{\gamma_{1,k}}} \right) + \frac{\frac{1+x_{2,0}\gamma_{2,n}}{\gamma_{1,n}}}{x_{1,0} + \frac{1+x_{2,0}\gamma_{2,n}}{\gamma_{1,n}}} + \sum_{m=1}^{M'} \left(\frac{\frac{1}{\gamma_{2,m}}}{x_{2,0} + \frac{1}{\gamma_{2,m}}} \right) + \frac{\frac{1+x_{1,0}\gamma_{1,n}}{\gamma_{2,n}}}{x_{2,0} + \frac{1+x_{1,0}\gamma_{1,n}}{\gamma_{2,n}}} \right] \quad (4.33)
$$

$$
\le \frac{1}{\ln 2} \left[\sum_{k=1}^{K'} \left(\frac{1}{x_{1,0}\gamma_{1,k} + 1} \right) + \sum_{m=1}^{M'} \left(\frac{1}{x_{2,0}\gamma_{2,m} + 1} \right) + 1 + \frac{1}{1 + x_{1,0}\gamma_{1,n} + x_{2,0}\gamma_{2,n}} \right]
$$

$$
= \frac{1}{\ln 2} \left[\sum_{k=1}^{K'} 2^{-b_k} + \sum_{m=1}^{M'} 2^{-b_m} + 1 + 2^{-b_n} \right]
$$

where $K'(K' \le K)$ and $M'(M' \le M)$ denote the number of subchannel with positive power allocation. b_j is the number of bits allocated in subchannel j.

Theorem 1 The constant-power joint-waterfilling can achieve a rate R' that is at most $\frac{1}{\ln 2}\left(\sum_{k=1}^{K'} 2^{-b_k} + \sum_{m=1}^{M'} 2^{-b_k} + 1 + 2^{-b_n}\right)$ away from the true joint-waterfilling, where the number of bits b_j allocated in subchannel j is given by $\log 2\left(1 + \gamma_{i,j}x_{i,j}\right)$.

The above theorem allows us to devise a low-complexity constant-power joint-waterfilling algorithm, as well as evaluate the conclusion numerically. Actually, the derived bound (4.33) is quite loose, for most cases the proposed scheme could achieve almost the same performance as true joint-waterfilling, as it shall be seen in the numerical result.

4.2.3 Performance Analysis and Evaluation

In this section, the performances of different power allocation strategies are compared by Monte-Carlo simulations. The sum rate of each power allocation scheme is averaged over 1,000 independent frequency-selective fading channel realizations. For the sake of simplicity, Path loss model adopted is Okumura-Hata: $l(d) = 13,774 + 35,22\lg(d)$ in dB, Shadowing's standard deviation is 3.65 dB, bandwidth is $B = 1$ MHz, and the frequency-selective channel is simulated to be block-fading with $N = 15$ subchannels. The downlink noise $N_0 \Delta B$ at each subchannel is assumed to be the same as -105 dBm. We also assume both the CTPs have the same power constraint as $P1 = P2 = P$.

Figure 4.6 compares the performance of constant-power joint-waterfilling with exact joint-waterfilling scheme and non-cooperative traditional waterfilling under different power constraint P (dBm). It is noted that the non-cooperative traditional waterfilling scheme here means each CTP separately allocates its power by traditional waterfilling solution. As expected, the exact joint-waterfilling and the proposed low-complexity scheme give indistinguishable results, both at low and high power constraint. Moreover, Fig. 4.6 also indicates that the cooperative power allocation schemes are always superior to non-cooperative traditional waterfilling, especially when the transmission power increases, similar conclusions has also been pointed out in many contributions [55, 56]. It is concluded that when there is no cooperation between the CTPs traditional, WF is just local optimal. A specific $P = 44$ (dBm) case is presented in Fig. 4.7 for convenient comparison.

Numerical simulation results show that the low-complexity constant-power joint-waterfilling has a negligible performance loss compared to true joint-waterfilling. Although a frequency-selective fading channel is mainly analyzed in this section, it is emphasized that the theoretical analysis can also be applied into cooperative OFDM systems which we are currently being utilized.

Fig. 4.6 Sum rate versus
power constraint P (dBm)

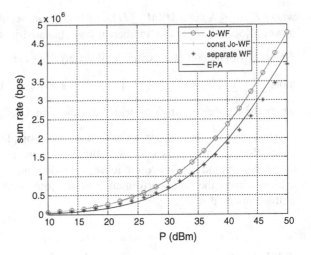

Fig. 4.7 Sum rate versus
power constraint $P = 44$
(dBm)

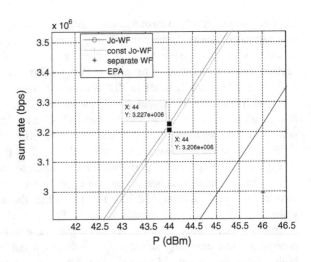

4.3 Summary

In this chapter, we focused on downlink power control for Group Cell systems.
Joint optimization subcarrier scheduling and power allocation scheme based on the
criterion of inter-cell interference coordination is proposed in Sect. 4.1. Simulation
results show that compared to fix subcarrier allocation and fixed power allocation,
the data rates at cell edge MTs are greatly increased by the proposed schemes,
which proves that such schemes with Group Cell can improve the performance at
cell edge. In Sect. 4.2 a novel power allocation scheme based on the criterion of
maximizing system capacity is proposed. In order to reduce the complexity of
transmitter and receiver design, a suboptimal power allocation scheme, constant

power allocation scheme, is presented and its performance lower bound is analyzed. Simulation results show that low-complexity constant-power joint-waterfilling has a negligible performance loss compared to true joint-waterfilling and both the schemes outperform non-cooperative traditional waterfilling scheme.

Chapter 5
Group Cell Trial Systems

In China, the research on the future mobile communication systems has been set up since 2001 [57, 58] The FuTURE program, which is supported by the National 863 Projects, aims at the development of the 4th generation (4G) mobile communication systems, and it includes two branches: FDD and TDD. As one of the two chief architects of the Chinese national 4G (4th generation) wireless communication system, supported by the Chinese Ministry of Science and Technology (MOST), the author led the Chinese 4G TDD working group comprising of four Chinese top-level universities, and established a Group Cell trial network adopted by the Chinese national 4G program in 2006 [59] mainly to test the performance of Group Cell architecture.

5.1 Introduction to FuTURE 4G TDD Trial System

5.1.1 Technical Targets

To demonstrate some features of Group Cell, 4G TDD Trial System supported by FuTURE program needs to meet some technical targets, which include reliable data transmission supporting wide dynamic range, high speed mobility and QoS requirements for different IP services under the different scenarios. The main technical targets in different scenarios are listed as following in Table 5.1.

5.1.2 Key Technologies and PHY Link Design

In order to realize the 100 Mbps peak data rate, a series of key PHY techniques [60–62], such as OFDM, MIMO, Turbo coding, high-order modulation synchro-

X. Tao et al., *Group Cell Architecture for Cooperative Communications*,
SpringerBriefs in Computer Science, DOI: 10.1007/978-1-4614-4319-3_5,
© The Author(s) 2012

Table 5.1 Scenarios for FuTURE 4G trial system

Scenarios	Spectrum efficiency (bps/ Hz)	BLER	E_b/N_0 (dB)	Test environment (km/h)
100 Mbps/cell (1 ~ 2 cells)	5	10^{-6}	3	5, 120, 250
2 × 50 Mbps/cell (1 cell)	5	10^{-6}	0	5, 120, 250
8 Kbps/cell (1 cell)	/	10^{-3}	0	5, 120, 250

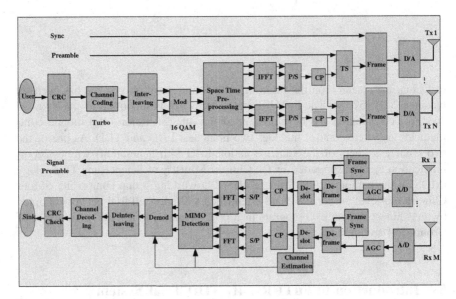

Fig. 5.1 The block structure of the FuTURE 4G TDD trial system

nization and channel estimation are applied into the FuTURE 4G TDD Trial System. The transmission link is shown in Fig. 5.1.

5.1.3 TDD Frame Structure

The radio frame structure [63] is depicted in Fig. 5.2. The duration of a radio frame is 5 ms, where TS0 is designed for the downlink dedicated signaling including system information broadcast and paging. A dedicated time slot (TS1) is designed for frequency synchronization in both uplink and downlink. The remaining slots are designed for data transmission for both uplink and downlink. The service slots can be designed in different structure so that it can flexibly support the resource allocation and reduce the overhead.

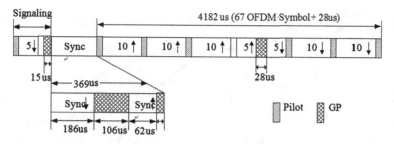

Fig. 5.2 Transmission frame structure

Table 5.2 The PHY parameter settings for FuTURE 4G TDD trial system	Parameters	Settings
	Frequency	3.5 GHz
	System bandwidth	20 MHz
	Subcarriers	1024
	Effective subcarriers	884
	Effective bandwidth	17.2656 MHz
	Subcarrier interval	19.5 kHz
	Cyclic extension	216(10.8 us)
	Symbol duration	51.2 + 10.8 = 62.0 us
	Modulation	16 QAM
	Turbo coding rate	1/3(after punching 1/2)
	Vehicular speed	5 ~ 250 km/h
	Channel model	COST207
	Antenna elements	4(MT) × 8(BS)

5.2 Performance Analysis and Evaluation

Various kinds of PHY key technologies are applied in the above system. To evaluate the performance of the whole link, various data services at the different scenarios are considered in simulation [64, 65]. The detailed parameter settings are shown in Table 5.2.

5.2.1 Simulation Scenarios

The considered simulation scenarios (including various data service with different vehicular speeds) are shown as followings [66]:

(1) *Uplink Performance of One User in Single-cell with* 8 Kbps *Service*
(2) *Uplink Performance of One User in Single-cell with* 100 Mbps *Service*
(3) *Uplink Performance of Two Users in Single-cell with* 50 Mbps *Service*
(4) *Uplink Performance of Two Users in Two-cell with* 100 Mbps *Service*

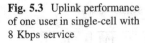

Fig. 5.3 Uplink performance of one user in single-cell with 8 Kbps service

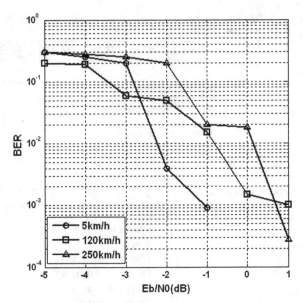

Besides, three kinds of vehicular speeds are considered in each case including 5, 120 and 250 km/h, which represents walking, medium and high vehicular speed separately. The two-user in two-cell (Group Cell) scenario is corresponding to multi-user MIMO scheme inside Group Cells as depicted in Fig. 1.6.

5.2.2 Performance Analysis and Evaluation

Simulation results on various scenarios transmission link in FuTURE 4G TDD Trial System are shown in Figs. 5.3–5.6.

It is proved by the above simulation results that the BER in FuTURE 4G TDD radio uplink transmissions is 10^{-3} when Eb/No is less than 1 dB, which satisfies the voice transmission requirement. Meanwhile, for high data rate services in single-user single-cell test scenarios and two-user single-cell test scenarios, the BER is 10^{-6} when Eb/No is less than 3 dB at different vehicle speeds. Above all, judging from the uplink system performance, the objectives of all the different services can meet the technical targets.

5.3 Trial Equipments and Trial Scenarios

The FuTURE 4G TDD Trial System adopts the uplink MIMO structure, which consists of mobile terminals (MTs) configured with an antenna unit (AU) and base stations (BSs) configured with two separated located AUs. And each AU consists

Fig. 5.4 Uplink performance of one user in single-cell with 100 Mbps service

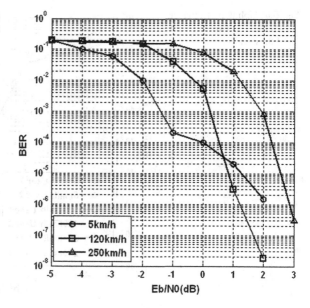

Fig. 5.5 Uplink performance of two users in single-cell with 50 Mbps service

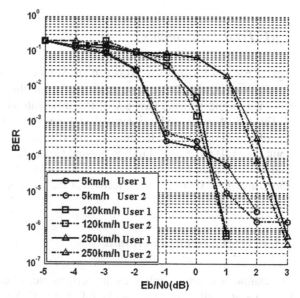

of four antenna elements (AEs). The transceiver can apply the adaptive $1 \times 1 \sim 4 \times 8$ MIMO structure, and meets the demands of various Group Cell system scenarios, which are described in Sect. 1.3 and 1.4 above.

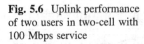

Fig. 5.6 Uplink performance of two users in two-cell with 100 Mbps service

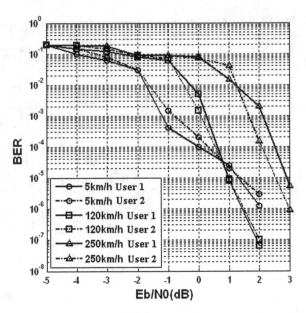

5.3.1 Trial Equipments

Figures 5.7 and 5.8 show two BSs and test vehicle of the FuTURE 4G TDD Trial System respectively. There are overall six AUs in the test field, and each AU can construct an independent Group Cell together with other AUs.

The two AUs of BS1 are located at Point A (AU1) and Point B (AU2) as described in Fig. 5.9, and the two AUs of BS2 are located at Point C (AU3) and Point D (AU4) as described in Fig. 5.9 and Guang Chang Building (AU5) and Bei Building (AU6) as described in Fig. 5.10.The AUs are connected to BSs by optical fibers, and the AUs of MTs are located at the roof-top of the test vehicle.

5.3.2 Trial Scenario: Campus

The campus scenario mainly tests the performance of coordinated communication inside Group Cell, which also can be referred to CoMP JP as depicted in Fig. 1.5. And hereafter we denote the scenario as JP. Table 5.3 shows the orientation parameters of distributed AUs in the trial scenario.

In order to test the Group Cell performance, we completed two comparative trials as shown in Figs. 5.9 and 5.10. In the campus trail scenario as depicted in Fig. 5.9, the northern base station (BS1) was composed of two remote four-antenna AUs, established on Hotel A (AU1) and Library (AU2), without supporting JP. The other southern (BS2), which has one Group Cell, that is composed of two remote

Fig. 5.7 Test vehicle of the
FuTURE 4G TDD trial
system

Fig. 5.8 BSs of the FuTURE
4G TDD trial system

four-antenna AUs, located on the Teaching building (AU4) and Hotel B (AU3),
which supports JP scenario inside the Group Cell. The distance of adjacent AUs is
about 200 m. The arrow headed red line is the routine of the test vehicle. (m, n;
$1 \leq m \leq 5$ and $1 \leq n \leq 10$) denotes the position of the mobile test vehicle.

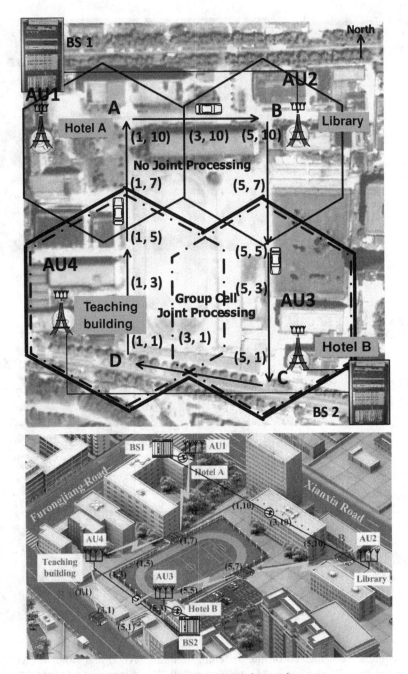

Fig. 5.9 JP of group cell field test scenario in typical urban environment

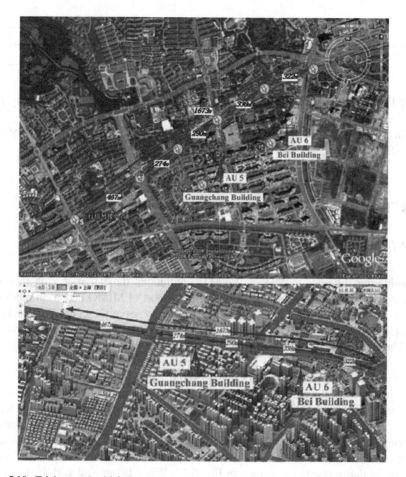

Fig. 5.10 Trial scenario: highway

Table 5.3 Orientation parameters of distributed AUs

AU	Horizontal plane	Down-tilt	Distance between AU (m)
D(AU4)	75° of southeast	15°	200
C(AU3)	75° of southwest	15°	200
A(AU1)	75° of northeast	15°	200
B(AU2)	75° of northwest	15°	200

5.3.3 Trial Scenario: Highway

The highway scenario mainly tests the performance of Slide Handover in Group Cell as depicted in Fig. 1.8, which is detailed discussed in Chap. 3. And the Slide Handover scenario is also corresponding to the CoMP scenario as depicted in

Table 5.4 Distance between each point on highway

/	S–A	A–B	B–C	C–D	D–E
Distance (m)	467	274	290	274	322

Fig. 1.10d. The out-field test map and the symbolic positions of Gubei highway are shown below.

In the figure, the distance between Point S and Point E is about 1673 m. More specifically, the distance is shown in Table 5.4.

5.4 Trial Results of Group Cell Trial System

5.4.1 Point-to-Point Link Performance Trial

Considering different mobile of the MT, continuous three frames data in time domain was obtained through sampling at the export of MIMO detection module. Then, by utilizing the mapping software, we can observe the constellation characteristics, obtained by the MIMO detection and then the performance is analyzed.

The frame structure is shown in Fig. 5.11, and the specific method using the pilot carriers for slot times of the routine business is shown in Fig. 5.11.

When MT moved slowly or with medium speed and the trial system was working in the uplink model with four transmitting antennas and eight receiving antennas, its actual MIMO detection results calculated are shown in Fig. 5.12. Maintaining the same speed, when there were frequency offsets at the multi-antenna radio frequency units belonging to the transmitter and the receiver of the trial system, which worked in the abnormal condition, the actual MIMO detection results of that scenario are shown in Fig. 5.13. When BS moved with high speed and still adopted the method using pilot carriers shown in Fig. 5.11, the actual MIMO detection results of the whole frame and each slot time are shown in Fig. 5.14.

When MT moved slowly or with medium speed, the detection results were precise and the constellation of the actual detection is very clear. On the other hand, when MT moved with high speed, the method using pilot carriers was causing amplitude differences and phase deviation of data symbols between the real value and the estimated value of the channel, so that if there were no frequency-offset estimation circuits, the actual MIMO detection results of each slot before and after pilot slot would have been in different constellation rotation directions [67, 68].

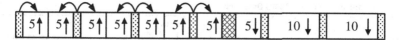

Fig. 5.11 Method using pilot carriers when executing the MIMO detection

Fig. 5.12 Constellation after MIMO detection working in the uplink (4 × 8)

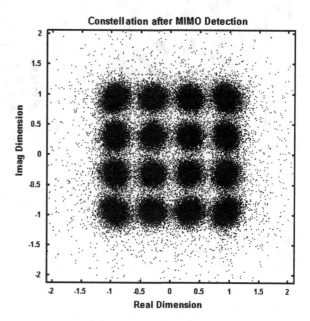

Fig. 5.13 Constellation after MIMO detection with frequency offsets

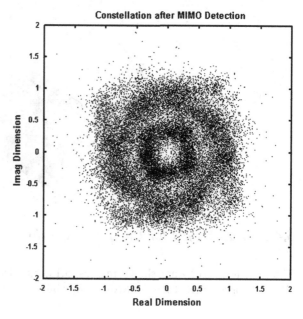

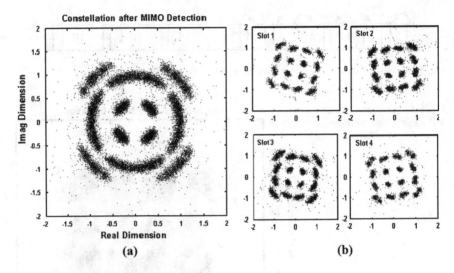

Fig. 5.14 Constellation rotating after MIMO detection. **a** Constellation of the whole frame after MIMO detection (*left*). **b** Constellation after MIMO detection of each slot time (*right*)

Fig. 5.15 BLER
performances of JP and non-
JP with 1×8 AEs

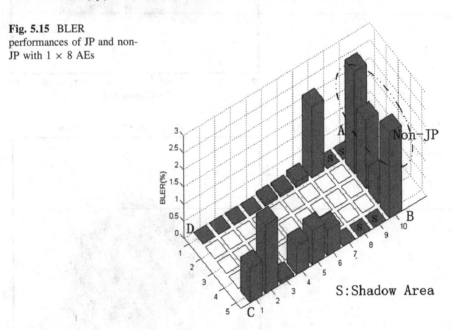

5.4.2 Trial Results for Campus Scenario

The field test results are shown in Figs. 5.15, 5.6, Tables 5.5 and 5.6 with the Block Error Rate (BLER) and SNR performance of JP and non-JP scenarios.

Table 5.5 SNR performances of JP and non-JP with 1×8 AEs

Test coordinate	SNR(dB) of BS2			SNR(dB) of BS1		
	AU3 (Hotel B)	AU4 (Teaching building)	BLER (%)	AU1 (Hotel A)	AU2 (Library)	BLER (%)
(1,1)	17,16,16,18	16,14,18,19	0.000000	/	/	/
(1,2)	16,18,19,15	14,17,14,16	0.000000	/	/	/
(1,3)	14,15,17,17	18,20,16,20	0.000000	/	/	/
(1,4)	17,15,19,20	13,15,11,17	0.000000	/	/	/
(1,5)	14,19,13,13	17,18,18,18	0.083333	/	/	/
(1,6)	14,12,7,12	18,17,19,18	0.083333	/	/	/
(1,7)	16,16,12,15	18,17,18,18	0.160000	/	/	/
(1,8)	10,13,7,10	16,18,16,18	2.083333	/	/	/
(1,9)[a]	0,4,0,0	12,16,16,16	78.41667	/	/	/
(1,10)[b]	/	/	/	16,16,18,17	12,13,14,12	29.45833
(2,10)	/	/	/	17,12,16,18	23,17,23,20	3.000000
(3,10)	/	/	/	19,20,20,20	16,15,20,17	2.000000
(4,10)	/	/	/	7,19,19,20	23,24,25,24	0.750008
(5,10)	/	/	/	18,18,19,20	19,18,26,23	2.500000
(5,9)[c]	/	/	/	0,4,7,4	15,15,16,11	33.58332
(5,8)[d]	10,18,9,17	0,13,7,9	17.00000	/	/	/
(5,7)	18,19,17,19	12,14,15,16	0.000000	/	/	/
(5,6)	17,18,12,19	15,13,18,17	0.750000	/	/	/
(5,5)	18,18,18,20	0,12,11,15	1.000000	/	/	/
(5,4)	17,14,18,17	7,13,14,14	0.833333	/	/	/
(5,3)	18,17,12,19	15,14,14,9	0.100000	/	/	/
(5,2)	17,17,18,18	16,12,19,17	2.000000	/	/	/
(5,1)	17,16,17,19	15,16,16,16	1.000000	/	/	/

[a, b, c, d] (1,9), (1,10), (5,8) and (5,9) are under cover of high buildings, thus the performance decreases.

Seen in Fig. 5.16, (2,10) ~ (5,10) is the non-CoMP zone near the northern BS1, which has bad performance of about 20 % BLER between Point A and Point B. The southern BS2 with Group Cell is well coordinated and supported by the CoMP JP scheme between Point C and Point D. The CoMP JP zone near Point C and Point D has good BLER performances inside of the Group Cell coverage. The field test BLER result in this region is below 0.1 %. Compared to the non-CoMP, the benefit of CoMP JP can be found obviously. Moreover, (1,7), (1,8) and (5,7) in the crossover zone of two BSs' edge are the handover area, which indicates quite bad performance of about 10 % BLER. (1,9), (1,10), (5,8) and (5,9) are the shadow area caused by the existing high building between Point A and Point B.

Table 5.6 SNR performances of JP and non-JP with 4 × 8 AEs

Test coordinate	SNR(dB) of BS2			SNR(dB) of BS1		
	AU4 (Teaching building)	AU3 (Hotel B)	BLER (%)	AU1 (Hotel A)	AU2 (Library)	BLER (%)
(5,1)	18,17,18,20	15,17,16,16	0.32	/	/	/
(5,2)	18,18,18,19	12,16,15,14	0.72	/	/	/
(5,3)	18,17,17,18	13,15,17,14	1.2	/	/	/
(5,4)	19,17,18,18	15,15,17,17	0.4	/	/	/
(5,5)	18,20,20,19	12,13,14,16	1.8	/	/	/
(5,6)	20,19,18,19	17,17,17,18	2.6	/	/	/
(5,7)	18,19,7,16	12,13,11,9	15.0	/	/	/
(5,8)	/	/	/	12,13,0,9	7,4,7,9	98.4
(5,9)	/	/	/	12,11,12,4	20,19,20,19	28.6
(5,10)	/	/	/	15,19,18,18	25,22,24,24	1.31
(4,10)	/	/	/	27,16,21,15	24,23,23,20	3.25
(3,10)	/	/	/	18,19,19,19	25,18,25,23	7.71
(2,10)	/	/	/	19,19,19,21	25,17,25,21	16.7
(1,10)	/	/	/	19,18,20,20	19,12,25,21	21.0
(1,9)	/	/	/	19,19,17,21	12,4,10,9	87.0
(1,8)	13,15,10,12	18,18,18,18	10.0	/	/	/
(1,7)	15,12,4,12	18,18,17,18	9.0	/	/	/
(1,6)	12,16,9,13	18,17,18,18	3.2	/	/	/
(1,5)	15,15,12,16	17,18,17,17	0.8	/	/	/
(1,4)	17,18,13,16	18,18,17,21	1.0	/	/	/
(1,3)	17,19,13,16	18,18,17,21	0.34	/	/	/
(1,2)	18,18,17,19	17,17,18,18	0.4	/	/	/
(1,1)	16,19,18,19	15,18,18,18	0.12	/	/	/

Fig. 5.16 BLER performances of JP and non-JP with 4 × 8 AEs

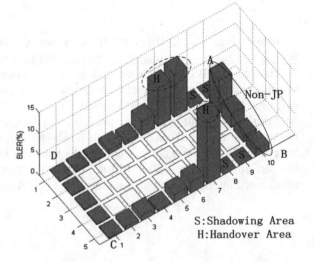

Table 5.7 Field test data of slide handover

Location of the test point	SNR(dB) of BS		BLER (%)
	AU5 (Guangchang building)	AU6 (Bei building)	
0 m(S)	10, 13, 4, 14	0, 0, 9, 13	2.416667
300 m	13, 14, 12, 14	13, 13, 15, 13	0.083333
467 m(A)	13, 14, 10, 13	13, 16, 16, 14	0.000000
558 m	12, 15, 12, 12	7, 15, 13, 13	0.250000
649 m	18, 15, 14, 9	12, 16, 17, 16	0.500000
741 m(B)	10, 14,12, 13	16, 17, 15, 14	0.083333
837 m	13, 17, 4, 15	15, 14, 13, 15	1.000000
933 m	14, 11, 13, 14	13, 15, 18,12	0.416667
1031 m(C)	12, 18, 16, 13	14, 15, 12, 19	0.333333
1122 m	12, 17, 13, 17	0, 16, 16, 10	1.416667
1213 m	15, 15, 13,15	16, 13, 12,11	0.166667
1305 m(D)	10, 13, 10, 15	14, 14, 13, 14	0.333333
1412 m	16, 13, 11, 18	15, 15, 12, 12	0.250000
1519 m	15, 18, 15, 17	0, 0, 0, 7	0.000000
1673 m(E)	18, 16, 14, 17	0, 0, 0, 4	0.000000
1723 m	16, 20, 16, 17	0, 0, 4, 0	0.000000
1773 m	15, 14, 13, 14	0, 0, 0, 0	2.583333

5.4.3 Trial Results for Highway Scenario

Highway Scenario is designed for Slide Handover in Group Cell.

Table 5.7 shows the field test data of Slide Handover from Point S to Point E with vehicular speed range from 20 to 60 km/h.

When the test vehicle moved from west to east on Gubei highway, one-way (uplink) link performance was tested for the FuTURE TDD system. The system maintained a good performance from Point A to Point E, and block error rate (BLER) was basically below 0.5 % and could even reached to zero in some certain positions. (Fig. 5.17)

For the moving test vehicle, the BLER occasionally increased by 1 % because of the change in surroundings and the intermittent shelter of some buildings, but it still met high data rate service requirements. Except the zone near Point E, Slide Handover and inter-cell cooperation frequently happened at most area, so it can be deduced that the system can also stay at low BLER, even if there exists the shelter of high buildings.

5.5 Trial Plan in the Next Phase

Since the relay channel was put forward by Cover in 1979, the cooperative ideas have been widely developed for over 30 years. In the year 2001, we proposed the concept of Group Cell, and accomplished the first Group Cell trial network in

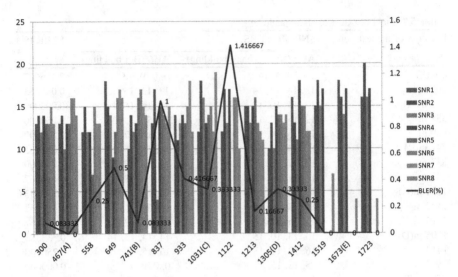

Fig. 5.17 Performance for slide handover

2006, which can be regarded as the first CoMP network, testing the performance of Joint Processing, Slide Handover and other cooperative technologies, although CoMP was presented officially in Apr. 2008. Results of a trial network implementation and a performance evaluation of the concept revealed that Group Cell can bring significant performance improvements. Besides, there were also some international projects focusing on CoMP research and field test, such as EASY-C project which is accomplished in 2009.

Table 5.8 shows the simulation parameters of Group Cell and implementation parameters of FuTURE 4G TDD Trial System at the first two columns and list the parameters of New Generation Broadband Mobile Communication Trial Network which we are going to work on at the last column.

For future evaluation and field test of cooperative communication techniques for "New Generation Broadband Mobile Communication Network", CoMP and Cooperative Relay based trial networks have been deploying inside the campus of Beijing University of Posts and Telecommunications since Jan. 2009 [70, 71], as shown in Fig. 5.18. The basic trial system parameters for field test are shown in Table 5.8.

Figure 5.18 presents detailed information of BUPT Trial Network. In the CoMP Trial Network scenario (from Mingguang Building to Baoweichu Building), new features of Coordinated Scheduling/Beamforming between two BSs will be evaluated and the performance of actual wireless environment will be demonstrated. In another Cooperative Relay Trial Network, consists of a BS at Main Teaching Building, a relay node at No.3 Teaching Building and a MT, Cooperative Relay technologies will be evaluated. New field test results and performance evaluation will be given in the following year. This work and the trial networks are supported by National Science and Technology Major Project "New Generation

Table 5.8 Parameters of group cell simulation and trial networks

Parameters	Group cell (CoMP) simulation	FuTURE 4G TDD trial	New generation broadband mobile communication trial network1
Carrier frequency	3.5 GHz	3.5 GHz	3.5, 5.8 GHz(for relay)
System bandwidth	20 MHz	20 MHz	20–100 MHz
Duplex mode	/	TDD	TDD
Peak data rate	100 Mbps	100–122 Mbps	100 M–1 Gbps
Transmission power	27 dBm	27 dBm	27 dBm
Active subcarrier/FFT point	880	884/1024	1664/2048
Subcarrier interval	15 kHz	19.5 kHz	60 kHz
Symbol period	/	62 us	18.75 us
Block size	/	4400 bits	3776 bits
Modulation	/	16QAM	16QAM
Channel coding	Turbo code	Turbo code/LDPC	LDPC
Antenna configuration(MT, BS)	(2, 4)	(1,1), (1,4), (2,4), (4,8)	(4,6)
Antenna separation(MT, BS) (in times of wavelength)	(10, 10)	(10, 10)	(10, 10)
Distance of adjacent AUs	200 m	200 m	200 m
Channel model	SCM, urban macro, high spread	/	/
MT speed	5 km/h	3 ~ 100 km/h	3 ~ 100 km/h
Distance-dependent path-loss	L = 128.1 + 37.6 log10(R), R in km	/	/
Shadow fading deviation	8 dB	/	/

(continued)

Table 5.8 (continued)

Parameters	Group cell (CoMP) simulation	FuTURE 4G TDD trial	New generation broadband mobile communication trial network1
Penetration loss	20 dB	/	/
Noise figure at MT	7 dB	7 dB	7 dB
Antenna gain	17 dBi for sector antenna	12 dBi for sector antenna	12 dBi for sector antenna
Traffic model	Full buffer	Full buffer, VoIP	Full buffer, VoIP
Link-to-system level mapping	Exponential effective SIR mapping (EESM)	/	/
BLER target	10 %	/	/
Receiver algorithm	Zero forcing (ZF)	Zero forcing (ZF)	Zero forcing (ZF)
Number of BS\AU \MT\Relay	19\57\570\0 (wrap-around, uniform in entire network)	3\6\2\0	3\4\3\1
Scenarios support	JP, CS/CB	BS1: non-JP; BS2: JP (named as adaptive MIMO in 2006 [69]	JP, CS/CB, cooperative relay

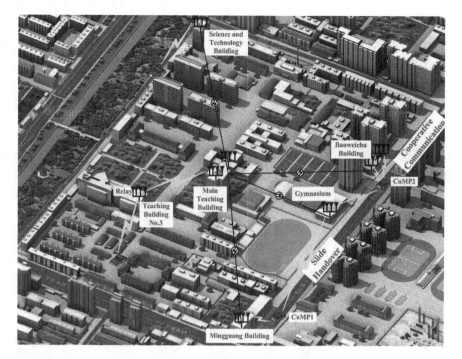

Fig. 5.18 BUPT trial networks for CoMP and cooperative relay based

Broadband Mobile Communication Network", which is the most important domestic program that aims at pushing forward the research, standard and industry innovations for 4G and beyond.

5.6 Summary

This chapter mainly introduced cooperative communication for the Group Cell system. Firstly, physical link design, the performance simulation evaluation, parameter design and scenario settings are introduced. Secondly, two typical networking environments of the Group Cell system are introduced, which are the campus scenario and the highway scenario. Finally, the actual performance is analyzed. Additionally, we also briefly introduced the future work.

References

1. X. Tao, P. Zhang, "Construction of Group Cell", China, Patent ZL 01137188.9,2001.11.8.
2. X. Tao, P. Zhang, "A novel wireless access network and the construction method", China, Patent ZL 01142473.7, 2011.11.29.
3. Bell Lab, "High-Capacity Mobile Telephone System Technical Report", Dec. 1971.
4. X.-H. You and X.-Q. Gao, "Development of Beyond 3G Techniques and Experiment System: An Introduction to the FuTURE Project", in ICT Shaping the World: A Scientific View, Wiley, Nov. 2008.
5. Thomas M. Cover and Abbas A. El Gamal, "Capacity Theorems for the Relay Channel", IEEE Transactions on Information Theory, IT, vol, 25, no.5, Sept. 1979, pp. 572–584.
6. A. Sendonaris, E. Erkip, and B. Aazhang, "Increasing uplink capacity via user cooperation diversity", in Proc. ISIT, Aug. 1998.
7. A.Saleh, A.Rustako , R.Roman, "Distributed Antennas for Indoor Radio Communications", Comm, IEEE Trans on, vol. 35, no. 12, Dec 1987, pp: 1245–1251.
8. H. Furukawa et al., "SSDT-Site Selection Diversity Transmission Power Control for CDMA Forward Link", IEEE Journal on Selected Area in Communications, vol. 18, no.8, Aug. 2000.
9. J. N. Laneman, G. W. Wornell, and D. N. C. Tse, "An efficient protocol for realizing cooperative diversity in wireless networks", in Proc. IEEE Int. Symp. Information Theory, Washington, DC, June 2001.
10. J. Nicholas Laneman and Gregory W. Wornell, "Distributed Space-Time-Coded Protocols for Exploiting Cooperative Diversity in Wireless Networks", IEEE Transactions on Information Theory, vol. 49, no. 10, Oct. 2003.
11. M. S. Alouini and A. J. Goldsmith, "Area spectral efficiency of cellular mobile radio systems", IEEE Transactions on Veh. Tech., vol. 48, no. 4, July 1999, pp. 1047–1066.
12. P. Zhang, X. Tao, et al, "A Vision from FuTURE Beyond 3G TDD", IEEE Communication Magazine, vol.43, no.1, Jan. 2005, pp. 38–44.
13. X. Xu, C. Wu, X. Tao, Y. Wang, P. Zhang, "Maximum Utility Principle Access Control for B3G Mobile System", Wireless Communications and Mobile Computing, Journal of Wiley, vol. 7, no. 8, Oct. 2007, pp. 951–959.
14. X. Tao et al., "Intelligent Group Handover Mode in Multicell Infrastructure", PIMRC 2003, Sept. 7–10, 2003, Beijing, China.
15. REV-080030, Ericsson, "LTE-Advanced – LTE evolution towards IMT-Advanced Technology components", 3GPP TSG RAN IMT Advanced Workshop Shenzhen, China, April 7–8, 2008.
16. GPP TS 36.211, V8.0.0 , "E-UTRA Physical channels and modulation", 3GPP, 2008.

17. X. Tao, Z. Dai, C. Tang, X. Xu, B. Liu, P. Zhang, "Capacity analyses for a generalized distributed antenna architecture for Beyond 3G systems", IEEE VTC 2005 Spring, May 29–June 1, Stockholm, Sweden.
18. Wonil Roh, Paulraj A, "Outage performance of the distributed antenna systems in a composite fading channel", IEEE VTC 2002 Fall, Vol 3, Sept. 2002, pp.24–28.
19. G. L. Stuber, *Principles of Mobile Communication*, Norwell, MA, Kluwer Academic Publisher, 1996.
20. Viswanath, P.; Tse, D.N.C.; Laroia, R, "Opportunistic beamforming using dumb antennas",Information Theory, IEEE Transactions on , Volume: 48 , Issue: 6 , June 2002 Pages:1277–1294.
21. Foschini G J, Gans M J, "On limits of wireless communications in a fading environment when using multiple antennas", Wireless Personal Communication, 1998, pp. 311–335.
22. D.N.C.Tse, "Optimal power allocation over parallel gaussian channels", Proc. of ISIT, 1997, p27.
23. Edelman A, Eigen values and condition numbers of random matrices. Cambridge, MA: Department of Mathematics, Massachusetts Institute of Technology, 1989.
24. 3GPP TSG-RAN R4-l01352 LTE-A "coexistence simulation assumptions".
25. Y. Wei, Lan T., "Transmitter Optimization for the Multi-Antenna Downlink with Per-Antenna Power Constraints", IEEE Trans.On Signal Processing,2007,55(6):2646–2660.
26. T. M. Cover and J. A. Thomas, *Elements of Information Theory*. New York: Wiley, 1991.
27. Indal N, Rhee W, Vishwanath S, et al, "Sum Power Iterative Power Iterative Water-Filling for Multi-Antenna Gaussian Broadcast Channels", IEEE Trans. On Information Theory, 2005,51(4).
28. S. Boyd, L. Vandenberghe. *Convex Optimization*. Cambridge, UK Cambridge University Press., 1985.
29. Y. Gong, X. Wang. "Channel Capacity Analysis and Simulations for Distributed MIMO System", IEEE Wicom' 2009: 1–4.
30. X. Tao, "Key Technologies for Advanced Mobile Communication Systems ", Ph.d thesis of Beijing University of Posts and Telecommunications, 2002
31. X. Xu and X. Tao, et al, "Fast cell group selection scheme for improving downlink cell edge performance", IEEE ICCCAS, Chengdu, 2006, (2): 1382–1386.
32. Harri Holma, Jaiine Laakso, "Uplink Admission Control and Soft Capacity with Mud in CDMA", IEEE VTC'99. pp431–435
33. Hiroshi Harada, Katsuyoshi Sato, Masayuki Fujise, "A Radio-on-Fiber Based Millimeter-Wave Road-Vehicle Communication System by a Code Division Multiplexing Radio Transmission Scheme", IEEE Transactions on Intelligent Transportation Systems, V01.2, No.4, Dec. 2001, pp165–179.
34. Alamouti, S.M., "A simple transmitter diversity scheme for wireless", IEEE Journal on Selected Areas in Communications, Vol. 16, Issue 8, Oct. 1998, pp. 1451–1458.
35. Foschini G. J. "Layered space-time architecture for wireless communication in a fading environment when using multi-element antennas", Bell Labs Technical Journal, 1996. pp. 41–59.
36. 3GPP, R1-050558, Motorola et al, "Simulation Assumptions and Evaluation for EUTRA".
37. 3GPP, R1-050448, Ericsson et al, "Basic principles for the Evolved UTRA radio access concept".
38. Rintamaki, M., Koivo, H., Hartimo, I., "Adaptive closed-loop power control algorithms for CDMA cellular communication systems", IEEE Transactions on Vehicular Technology, 2004.
39. Gerakoulis D., Salmi P., "An interference suppressing OFDM system for ultra wide bandwidth radio channels", IEEE Conference on Ultra Wideband Systems and Technologies, 2002.
40. 3GPP R1-083115. "Discussion on DL coordinated multipoint transmission", August, 2008.

41. Venkataraman V., Shynk J.J., "Adaptive interference suppression in multiuser OFDM", IEEE 60th VTC2004-Fall, 2004.

42. Hui Zhang, Xiaodong Xu, Jingya Li, Xiaofeng Tao, "Subcarriers Resource Optimization for Cooperated Multipoint Transmission", International Journal of Distributed Sensor Networks, 2010.

43. L. Gao, S. Cui, F. Li, "CTH01-3: A Low-complexity Adaptive Subcarrier, Bit, and Power Allocation Algorithm for OFDMA Systems", IEEE Global Telecommunications Conference, 2006.

44. J. Wu, et al, "On performance of multiple access for OFDM transmission technique in Rayleigh fading channel", International Conference on Communication Technology, vol.2, pp: 1025–1028, Apr. 2003.

45. S.Pietrzyk, et al, "Radio resource allocation for cellular networks based on OFDMA with QoS guarantees", IEEE GLOBECOM, vol.4, pp: 2694–2699, Dec. 2004.

46. 3GPP TR 25.814. "Physical layer aspects for evolved UTRA", October 2005.

47. G.J. Foschini, et al, "A simple distributed autonomous power control algorithm and its convergence", IEEE Trans. on Vehicular Technology. no.42, vol.4, pp. 641–646, 1993.

48. 3GPP R1-050896, "Description and simulations of interference management technique for OFDMA based E-UTRA downlink evaluation", 2005.

49. 3GPP TR 23.882, "System architecture evolution report on technical options and conclusions", 2006.

50. Bing Luo, Qimei Cui, Hui Wang, Xiaofeng Tao, "Optimal Joint Water-filling for OFDM Systems with Multiple Cooperative Power Sources", the IEEE Global Communications Conference (GLOBECOM) 2010, Dec. 2010,Miami, Florida,USA.

51. Fletcher R., Practical methods of optimization, Vol.1: Unconstrained optimization. New York: John Wiley & Sons, 1980.

52. Y. Wei, Cioffi, J.M., "Constant-power waterfilling: performance bound and low-complexity implementation", IEEE Transactions on Communications, 2006.

53. P. S. Chow, "Bandwidth optimized digital transmission techniques for spectrally shaped channels with impulse noise", 1993, Stanford University.

54. P. Viswanath, D. N. C. Tse, and V. Anantharam, "Asymptotically optimal water-filling in vector multiple-access channels", IEEE Trans. Inform.Theory, vol. 47, no. 1, pp. 241–267, Jan. 2001.

55. Hyang-Won Lee and Song Chong, "Downlink resource allocation in multi-carrier systems: frequency-selective vs. equal power allocation", IEEE Trans. on Wireless Comm., vol.7, no.10, pp.3738–3747, Oct. 2008.

56. X. You, G. Chen, M. Chen, et al, "The FuTURE Project in China", IEEE Communications Magazine, Vol. 43, Issue 1, 2005, pp. 70–75.

57. http://www.863.org.cn/863_105

58. P. Zhang, X. Tao, J. Zhang, et al, "The Visions from FuTURE Beyond 3G TDD", IEEE Communications Magazine, Vol. 43, Issue 1, 2005, pp. 38–44.

59. Bingham J.A.C., "Multicarrier Modulation for Data Transmission: An Idea Whose Time Has Come", IEEE Comm. Magazine, Vol. 28, Issue 5, May 1990.

60. Foschini G. J., Gans M. J., "On Limits of Wireless Communications in a Fading Environment when Using Multiple Antennas", Wireless Personal Communications, Vol. 6, Issue 3, 1998, pp. 311–335.

61. J. Jiang, P. Zhang, M. Zhou, L. Li, X. Tao, X. Ji, "A method of MIMO-OFDM system channel estimation". China, Patent No.CN1917397, 2006.9.19.

62. L. Li, M. Zhou, X. Tao, et al, "Adaptive Frame Structure in B3G-TDD Uplink", In Wireless Communications and Mobile Computing Special Issue on Asia-Pacific B3G R&D Activities and Technology Innovations, Vol. 7, Issue 8, Oct. 2007, pp. 985–993.

63. Michael D. Ciletti, "Advanced Digital Design with the Verilog HDL", Publishing House of electronics industry, 2005, pp. 533–556.

64. Booth AD., "A Signed Binary Multiplication Technique", Quarterly Journey of Mechanics and Applied Mathematics, 1951.
65. X. Tao, Z. Dao, C. Tang, et al, "General Cell Architecture and Handover–Group cell and Group Handover", Journal of Electronics Dec, 2004,Vol. 32, pp. 114–117.
66. X. Tao, Z. Yu, H. Qin, et al, "New Sub-optimal Detection Algorithm of Layered Space-time Code", In VTC'02, Vol. 4, May 2002, pp. 1791–1794.
67. X. Tao, Z. Yu, H. Qin, et al, "Sub-optimal Decoding Algorithm of V-BLAST", Journal of Electronics May, 2003,Vol. 31, pp. 703–706.
68. X. Tao, Costa Elena , Z. Yu, et al, "New Detection Algorithm of V-BLAST Space-time Code", In VTC'01, Oct. 2001, Vol. 4, pp. 2421–2423.
69. X. Tao, P. Zhang, J. Xu, et al, "An Antenna Imbalance Protection Method of the MIMO System", China. Patent No.200610153013.8
70. J. Xu, X. Tao, P. Zhang, "FuTURE 4G TDD Trial System and Field Test", In Global Mobile Congress(GMC'2007), Shanghai, China, Oct. 2007, pp. 124–129.